T0233975

Lecture Notes in Computer Science 11820

More information about this series at http://www.springer.com/series/8183

Marina L. Gavrilova · C. J. Kenneth Tan (Eds.)

Transactions on Computational Science XXXIV

Editors-in-Chief
Marina L. Gavrilova
Department of Computer Science
University of Calgary
Calgary, AB, Canada

C. J. Kenneth Tan
Sardina Systems OÜ
Tallinn, Estonia

ISSN 0302-9743 ISSN 1611-3349 (electronic)
Lecture Notes in Computer Science
ISSN 1866-4733 ISSN 1866-4741 (electronic)
Transactions on Computational Science
ISBN 978-3-662-59957-0 ISBN 978-3-662-59958-7 (eBook)
https://doi.org/10.1007/978-3-662-59958-7

This Springer imprint is published by the registered company Springer-Verlag GmbH, DE
part of Springer Nature
The registered company address is: Heidelberger Platz 3, 14197 Berlin, Germany

LNCS Transactions on Computational Science

Computational science, an emerging and increasingly vital field, is now widely recognized as an integral part of scientific and technical investigations, affecting researchers and practitioners in areas ranging from aerospace and automotive research to biochemistry, electronics, geosciences, mathematics, and physics. Computer systems research and the exploitation of applied research naturally complement each other. The increased complexity of many challenges in computational science demands the use of supercomputing, parallel processing, sophisticated algorithms, and advanced system software and architecture. It is therefore invaluable to have input by systems research experts in applied computational science research.

Transactions on Computational Science focuses on original high-quality research in the realm of computational science in parallel and distributed environments, also encompassing the underlying theoretical foundations and the applications of large-scale computation.

The journal offers practitioners and researchers the opportunity to share computational techniques and solutions in this area, to identify new issues, and to shape future directions for research, and it enables industrial users to apply leading-edge, large-scale, high-performance computational methods.

In addition to addressing various research and application issues, the journal aims to present material that is validated – crucial to the application and advancement of the research conducted in academic and industrial settings. In this spirit, the journal focuses on publications that present results and computational techniques that are verifiable.

Scope

The scope of the journal includes, but is not limited to, the following computational methods and applications:

- Aeronautics and Aerospace
- Astrophysics
- Big Data Analytics
- Bioinformatics
- Biometric Technologies
- Climate and Weather Modeling
- Communication and Data Networks
- Compilers and Operating Systems
- Computer Graphics
- Computational Biology
- Computational Chemistry
- Computational Finance and Econometrics

- Computational Fluid Dynamics
- Computational Geometry
- Computational Number Theory
- Data Representation and Storage
- Data Mining and Data Warehousing
- Information and Online Security
- Grid Computing
- Hardware/Software Co-design
- High-Performance Computing
- Image and Video Processing
- Information Systems
- Information Retrieval
- Modeling and Simulations
- Mobile Computing
- Numerical and Scientific Computing
- Parallel and Distributed Computing
- Robotics and Navigation
- Supercomputing
- System-on-Chip Design and Engineering
- Virtual Reality and Cyberworlds
- Visualization

Editorial

The *Transactions on Computational Science* journal is published as part of the Springer series *Lecture Notes in Computer Science*, and is devoted to a range of computational science issues, from theoretical aspects to application-dependent studies and the validation of emerging technologies.

The journal focuses on original high-quality research in the realm of computational science in parallel and distributed environments, encompassing the theoretical foundations and the applications of large-scale computations and massive data processing. Practitioners and researchers share computational techniques and solutions in the field, identify new issues, and shape future directions for research, as well as enable industrial users to apply the presented techniques.

The current issue is devoted to research on data analytics using machine learning and pattern recognition, with applications in wireless networks, databases, and remotely sensed data. It is comprised of seven regular papers.

The first paper introduces a machine learning method to extract meaningful signatures from the unstructured data using a hybrid approach. The research is validated by a collection of satellite images and establishes superiority of the method over traditional approaches for identification of important information from remotely sensed data.

The second article continues the topic of data mining, in the context of Knowledge Discovery in Databases (KDD), and presents a new method for discovering interesting, valid, meaningful, and previously unknown patterns from a large amount of data using a grid-based clustering for aggregate pattern mining.

The third article is an extended journal version of a conference paper that appeared in the proceedings of the International Conference on Computational Science and its Applications in 2016. It presents a comprehensive study of three different approaches to point placement on a line problem in an inexact model.

The forth article continues the topic of geometric processing and presents a new library that could be highly useful for researchers developing meshing algorithms and analysis tools for general polygonal and polyhedral meshes.

The fifth article presents a mechanism for trust computation in VANET cloud. The proposed algorithm takes into consideration the uncertainty and fuzziness associate with trust values by incorporating DST (Dempster Shafer Theory) and fuzzy analyzer.

The sixth paper of the issue demonstrates the effectiveness of cooperative communication approach for a Wireless Sensor Network (WSN) deployed in a snowy environment.

The final paper presents a built-in self-reconfiguring system for mesh-connected processor arrays, where faulty processing elements are compensated for by spare processing elements located on the diagonal. It is effective in enhancing the run-time reliability of a processor array in mission critical systems.

We thank all of the reviewers for their diligence in making recommendations and evaluating revised versions of the papers presented in this TCS journal issue. We would also like to thank all of the authors for submitting their papers to the journal and the associate editors for their valuable work.

It is our hope that this collection of seven articles presented in this special issue will be a valuable resource for *Transactions on Computational Science* readers and will stimulate further research in the vibrant field of computational science theory and applications.

June 2019

Marina L. Gavrilova
C. J. Kenneth Tan

LNCS Transactions on Computational Science – Editorial Board

Contents

Machine Learning in Hybrid Environment for Information Identification with Remotely Sensed Image Data

Rik Das[1(✉)], Sourav De[2], and Sudeep Thepade[3]

[1] Department of Information Technology, Xavier Institute of Social Service, Ranchi, Jharkhand, India
rikdas78@gmail.com

[2] Department of Computer Science and Engineering, Cooch Behar Government Engineering College, Cooch Behar, India
dr.sourav.de79@gmail.com

[3] Department of Computer Engineering, Pimpri Chinchwad College of Engineering, Pune, India
sudeepthepade@gmail.com

Abstract. Multi sensor image data used in diverse applications for Earth observation has portrayed immense potential as a resourceful foundation of information in current context. The scenario has kindled the requirement for efficient content-based image identification from the archived image databases to provide increased insight to the remote sensing platform. Machine learning is the buzzword for contemporary data driven decision making in the domain of emerging trends in computer science. Diverse applications of machine learning have exhibited promising outcomes in recent times in the areas of autonomous vehicles, natural language processing, computer vision and web searching. An important application of machine learning is to extract meaningful signatures from the unstructured data. The process facilitates identification of important information in the hour of need. In this work, the authors have explored the application of machine learning for content based image classification with remotely sensed image data. A hybrid approach of machine learning is implemented in this work for enhancing the classification accuracy and to use classification as a pre cursor of retrieval. Further, the approaches are compared with respect to their classification performances. Observed results have revealed the superiority of the hybrid approach of classification over the individual classification results. The feature extraction techniques proposed in this work have ensured higher accuracy compared to state-of-the-art feature extraction techniques.

Keywords: Remote sensing · Machine learning · Earth observation · Image classification · Information fusion

M. L. Gavrilova and C. J. K. Tan (Eds.): Trans. on Comput. Sci. XXXIV, LNCS 11820, pp. 1–28, 2019.
https://doi.org/10.1007/978-3-662-59958-7_1

1 Introduction

The significance of machine learning has positively impacted the growing use of satellite imagery for remotely sensed image data. Conventional text based annotation of image data stored in massive image datasets has several limitations. Nevertheless, the rising popularity of machine learning techniques has paved the way to design fruitful algorithms based on image content. Classification of images is a supervised technique which is dependent on pre-conceived knowledge of the classifier by means of training data. Training data for classification based on image content can be devised by exploring varied low level image features which include colours, shape, texture etc. of an image. Application of machine learning techniques on remotely sensed images has provided crucial information for diverse initiatives including forecasting of catastrophe, surveillance of environment, ecological survey, and so on [1, 2]. Increasing demand for remotely sensed images has necessitated the launching of numerous satellites for daily acquisition of thousands of images [3]. Traditional means of information identification with images are based on text based recognition system. The process maps images with text based annotations known as keywords which comprises of geographic location, sensor type, and time of acquisition. However, text or keywords based mapping of images has insufficient information about image contents and is based on perceived vocabulary of the individual annotating the images [4]. These are major drawbacks for recognition of information with text based annotation for images and the process refrains escalating the success rate with irrelevant outputs. Increased use of high end image capturing devices for remote sensing has yielded images with higher resolutions, clarity and enriched features for content based image classification. These features can be readily extracted for classification using popular deep learning techniques to represent the image contents without considering the entire image as input. However, deep learning techniques are computationally expensive and time consuming [5]. Other benchmarked techniques for feature extraction, namely, Uniform Local Binary Pattern (uLBP) [6], color histogram [7] and so on mostly extract features with higher dimensions which have computational overhead for processing during real time applications. Increased size of features enhances the convergence time of the classification algorithm and attracts high computational complexity [8]. Consequently, it is essential to design feature vectors with reduced dimension so that they can be used for faster real time decision making.

Therefore, the motivation of this work cán be enlisted as follows:

- Drawbacks of text based annotation for image classification
- Computationally expensive deep learning techniques
- Enhanced convergence time for classification results due to large feature size.

In this work, the authors have attempted to fuse the richness of smaller dimension extracted features by using two different extracted descriptors. The first technique involves selection of threshold locally using Feng's algorithm [9] and the second one uses Bottom Hat Morphological Transform for extracting signatures [10]. The authors have attempted to design a hybrid environment for classification. The environment is referred as hybrid due to its attempt to amalgamate diverse techniques for achieving single goal of enhanced classification. The hybrid environment is nurtured by implementing fusion based approach. The fusion technique is inspired by a popular method in machine learning, named Z score normalization. It is used to restrict dissimilar variables to dominate each other. Finally, the features extracted with two different techniques are standardized so that the classification results are not biased towards the features with higher "weights". Primarily, the aforesaid techniques are tested for individual classification accuracies and further both of them are compared to the fusion based hybrid techniques.

Thus, the objectives of this work are enlisted below:

- Extraction of feature vectors by reducing their dimensions.
- Facilitating early and late fusion for classification process.
- Comparison of the results to the benchmarked techniques.

The organization of the paper is as follows. Introduction of the topic is followed by review of Related Work. Our Approach is discussed in the subsequent section. Further, the description of Experimental Setup is elaborated which is implemented to calculate and portray Results and Discussions. Final section of the paper is Conclusion.

2 Related Work

Our attempt is to develop a hybrid machine learning architecture to facilitate content based image identification with remotely sensed image data. It has endeavored to combine the classification decisions obtained with separately extracted image features to improve the classification accuracy. Thereafter, the classification process is implemented to categorize the input of a retrieval process to reduce the computational overhead of searching during retrieval. The feature extraction techniques discussed in this work are from the domain of image binarization and morphology. Thus the literature survey embraces these two aforesaid domain and also includes fusion based machine learning since the work has implemented a hybrid machine learning architecture.

Extracting Descriptors with Binarization of Image

Selection of threshold globally, applying Otsu's algorithm has extracted the essential features for ground objects in remotely sensed images [11]. Adaptive threshold based technique is exploited for identifying feature vectors in remote sensing [12]. Image binarization technique has significant dependence on assortment of appropriate threshold. Nonetheless, there are varied components to influence the computation of threshold for binarization process unfavorably which comprise of patchy luminosity, insufficient contrast etc. [13]. Therefore, researchers have developed threshold selection techniques taking into account mean and standard deviation of the gray values locally and globally within an image, instead of considering only the mean of all gray values. Deployment of mean threshold selection with single level of threshold computation and multiple level of threshold computation have resulted in augmented classification outcomes [14]. However, the stretch of gray values within the image content is not properly addressed by mere selection of mean threshold which results in non inclusion of important details in determining image features for the classification process. This has resulted in consideration of standard deviation in designing threshold selection techniques globally and locally [15–17]. Extracted features using the aforesaid global and local threshold selection techniques have significantly contributed in augmented accuracy for classification process.

Extracting Descriptors with Morphological Operators
Edge detectors are used for feature extraction to classify the panchromatic high resolution statistics from the urban area [18]. Smallest Uni-value Segment Assimilating Nucleus (SUSAN) is used to detect edges for extraction of useful features by calculating the moment invariants of the edge map in remotely sensed image data [19]. Seashore boundary detection in remotely sensed images is successfully executed with morphological approach [20]. The researchers have extensively carried out experiments using popular shape descriptors based on contour and region operators which have attracted extensive coverage in recent literature [21, 22]. Boundary lines are highlighted by the application of morphological feature extraction based on contour. The function of highlighting the boundary lines by contour descriptors are based on well known mathematical descriptors based Fourier transform [23], curvilinear contours [24], and chain codes [25]. Extraction of significant information from composite shapes is effectively implemented using region-based descriptors, useful for feature extraction embracing entire object area [26].

Fusion Methodologies and Multi Technique Feature Extraction
Improvement of existing fusion techniques are carried out recently by the researchers by leveraging the restricted intensity of the process called hue saturation transform which has created hybrid of bands by simultaneous inclusion of three or more in number [27], by increasing the spectral veracity of the IHS blended image [28], by conducting the frequency domain amalgamation with a Fast Fourier Transform (FFT) [29], or by incorporating an analysis involving diverse resolution (MRA), such as the use of principal component substitution (PCS) by jointly performing wavelet transform (WT) with a principal component analysis (PCA) [30]. Four different varieties of data fusion is evident in recent narratives which are carried out by fusing data at an early stage, later stage, hybrid approach and intermediate approach. Early fusion typically integrates different features from various techniques and represents them as a single learner input. Dimension of the feature thus created is typically elevated due to blending of a range of values. Late fusion process treats the different features extracted with diverse techniques separately for different learners and finally uses a data standardization technique to fuse the classification decision. The correlations related to feature level can hardly be explored with this technique in-spite of its scalability compared to the earlier technique. However, both the aforesaid methods can be implemented together with hybrid fusion model. The last one is known as intermediate fusion which unites manifold features with a mutual model which results in increased precision [31]. Modern researchers have adopted more multifaceted approaches depending on the attributes like spectral content, number of bands, and have amalgamated various methods to experiment fusion based outcomes [32–36].

A common drawback with the techniques discussed in the contemporary literature is extraction of hefty feature vectors which have resulted in sluggish image identification process. The authors have addressed the limitation and have proposed techniques to extract feature vectors with considerably smaller dimension which is independent of image size. The fusion based machine learning architecture for recognition of images based on content using fresh perspective of feature extraction has revealed superior performance compared to existing techniques and has exhibited significance with respect to individual techniques.

3 Our Approach

In this approach, feature extraction is performed twice. One set of feature is extracted with Feng's method which is a recognized technique for local threshold selection to binarize image data for feature extraction. Another set of feature is extracted with morphological operator for shape feature extraction. The multi technique feature extraction process has explored the richness of feature set hidden in the content of the images which can contribute towards superior classification and retrieval results by instilling fusion framework. Elaborate discussions on the techniques for feature extraction are covered in the subsequent subsections along with the architecture for fusion.

Feature Extraction with Image Binarization

Usually the process of binarization is based on computation of threshold followed by comparison of pixel values to the threshold. The comparison process results in two groups of pixel data assigned with a value of 1 or 0. Hence, the image is to be divided into two different classes of foreground and background pixels based on the threshold value of binarization and representing the classes by 1's and 0's respectively. Thus, the feature size has the equivalent dimension to that of the image size as the entire image is represented in binarized fashion and the size of feature would be dependent on image size. In our approach, the two different groups formed by grey values corresponding to 1 and 0 respectively are computed for their subsequent mean and standard deviation values. The mean of the cluster added to the corresponding standard deviation of the cluster of pixel values assigned to 1 are considered as higher intensity group. We have also considered the addition of mean with standard deviation of the other cluster of pixel values assigned to 0 and considered it as lower intensity group. A given image is separated into three color components namely Red (R), Green (G) and Blue (B) respectively. Thus, the feature size is reduced to 2 for each color component in an image. Feng's threshold selection technique is used to select the local threshold for binarization [37]. Image contrast has influenced the formulation of the decision for binarization. Initially, the technique is governed by calculation of local mean (m) minimum (M) and the standard deviation (s) by shifting a primary local window. The effect of illumination is compensated by computing the dynamic of standard deviation (R_s) for a larger local window named as the secondary local window as shown in Fig. 1.

Exact size of the local window is decided by the illumination deviation while setting up the image capturing equipment. The threshold value is calculated as in Eq. 1.

$$T = (1 - \alpha_1) \bullet m + \alpha_2 \bullet (\frac{s}{R_s}) \bullet (m - M) + \alpha_3 \bullet M \tag{1}$$

where,

$$\alpha_2 = k_1 \bullet (\frac{s}{R_s})^{\gamma} \text{and} \alpha_3 = k_2 \bullet (\frac{s}{R_s})^{\gamma},$$

where α_1, γ, k_1 and k_2 are positive constants.

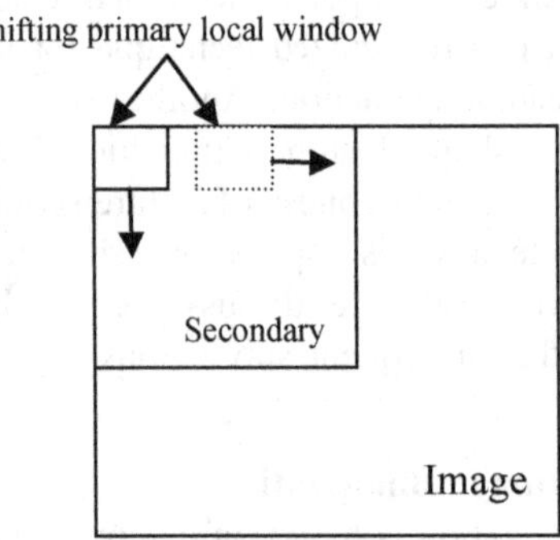

Fig. 1. Process of local threshold selection with Feng's method

The equation has three corresponding coefficients α_1, α_2 and α_3 for three elements to administer higher compliance and adaptability on the weight of diverse elements of equation. Formulations of α_2 and α_3 are carried out adaptively based on normalized local standard deviation (*s/Rs*) as it is assumed that windows having the foreground image has larger standard deviation compared to the background. Hence, inclusion of (*s/Rs*) ratio in coefficients α_2 and α_3 has permitted the computed threshold to separate the foreground and the background image efficiently without much subsequent knowledge of the input image. The consequence of binarization applying Feng's local threshold selection is illustrated in Fig. 2.

The gray values are compared to the corresponding threshold values and two clusters were formed comprising of gray values higher than the corresponding local threshold and gray values lower than the corresponding local threshold. Computation

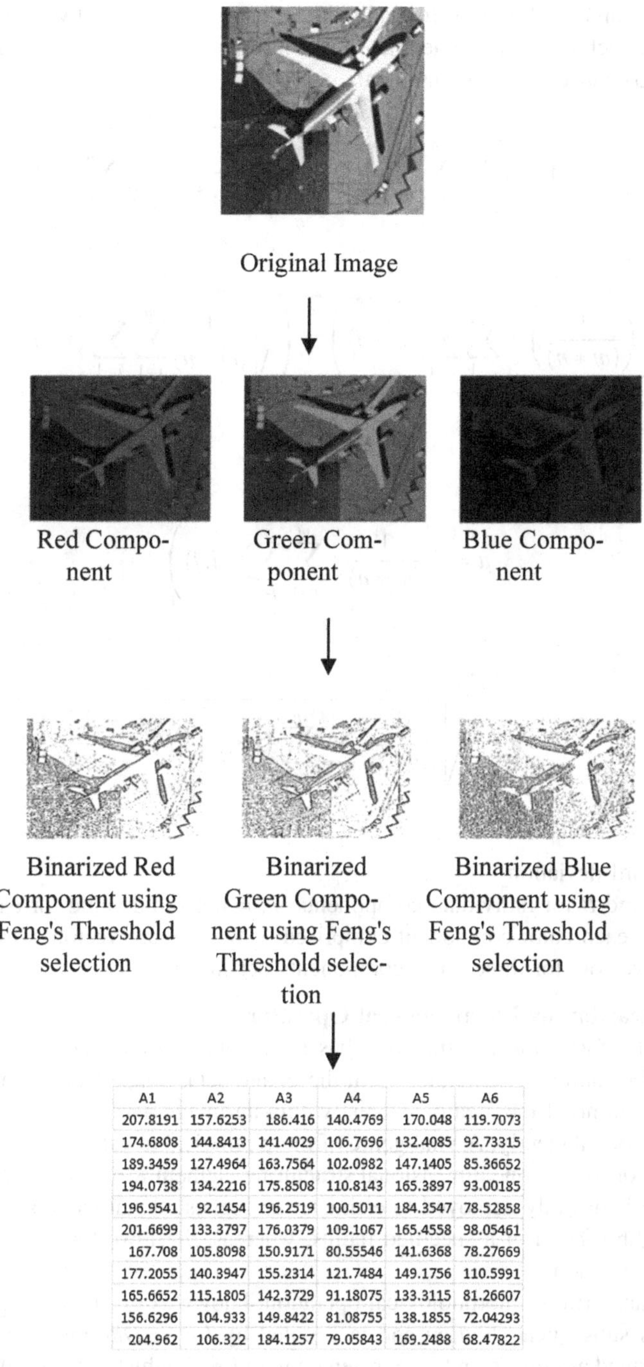

A1	A2	A3	A4	A5	A6
207.8191	157.6253	186.416	140.4769	170.048	119.7073
174.6808	144.8413	141.4029	106.7696	132.4085	92.73315
189.3459	127.4964	163.7564	102.0982	147.1405	85.36652
194.0738	134.2216	175.8508	110.8143	165.3897	93.00185
196.9541	92.1454	196.2519	100.5011	184.3547	78.52858
201.6699	133.3797	176.0379	109.1067	165.4558	98.05461
167.708	105.8098	150.9171	80.55546	141.6368	78.27669
177.2055	140.3947	155.2314	121.7484	149.1756	110.5991
165.6652	115.1805	142.3729	91.18075	133.3115	81.26607
156.6296	104.933	149.8422	81.08755	138.1855	72.04293
204.962	106.322	184.1257	79.05843	169.2488	68.47822

Extracted Feature vectors of dimension 6 per image for 11 images

Fig. 2. Process of feature extraction with binarization (Color figure online)

of mean and standard deviation for individual cluster has derived two feature vectors x_{hi} and x_{lo} for each color component as shown in Eqs. 2 and 3, where x stands for Red (R), Green (G) and Blue (B) components.

$$x_{hi\,F.V.} = \left(\left(\frac{1}{(m*n)}\right)*\right)\left(\sum_{i=1}^{m}\sum_{j=1}^{n} x_{hi}(i,j)\right) + \left(\sqrt{\frac{1}{(m*n)}\sum_{i=1}^{m}\sum_{j=1}^{n}(x_{hi}(i,j)-\mu_{hi})^2)}\right) \tag{2}$$

$$x_{lo\,F.V.} = \left(\left(\frac{1}{(m*n)}\right)*\sum_{i=1}^{m}\sum_{j=1}^{n} x_{lo}(i,j)\right) + \left(\sqrt{\frac{1}{(m*n)}\sum_{i=1}^{m}\sum_{j=1}^{n}(x_{lo}(i,j)-\mu_{lo})^2}\right) \tag{3}$$

where,

$$\mu = \left(\frac{1}{(m*n)}*\sum_{i=1}^{m}\sum_{j=1}^{n} x(i,j)\right)$$

And

$$\sigma = (\sqrt{\frac{1}{(m*n)}\sum_{i=1}^{m}\sum_{j=1}^{n}(x(i,j)-\mu)^2)}$$

μ = mean
σ = standard deviation
$x = R$, G and B for individual components, T_x = threshold value for each pixel

Therefore, each color component comprised of two feature vectors which sums up to 6 feature vectors for three color components per image.

Feature Extraction by Morphological Operator

Object based information identification has remarkably utilized the shape feature for extraction of meaningful signatures from the image data. Shape feature can efficiently identify the outline of the area of interest within an image and thus carry out contour based extraction efficiently. A vital genre of morphological transform is represented by the *hat*-transforms. If a one dimensional structuring element K of a signal f is assumed, then *dilation* is usefully deployed to estimate the highest signal value spread across a circular neighbor-hood of a specified radius, whereas, the lowest signal value can be computed by erosion.

A vital transform for morphological operation, known as *opening* is represented by *erosion* with subsequent *dilation* and is denoted by $f \circ K$. *Dilation* with subsequent *erosion* is termed as *closing* and is represented with $f \bullet K$, which is the dual of *opening*. *Top hat* transform is described as the residual of the *opening* in contrast to the original signal and is denoted by $f - (f \circ K)$. The bottom hat transform, known as, dual of *Top*

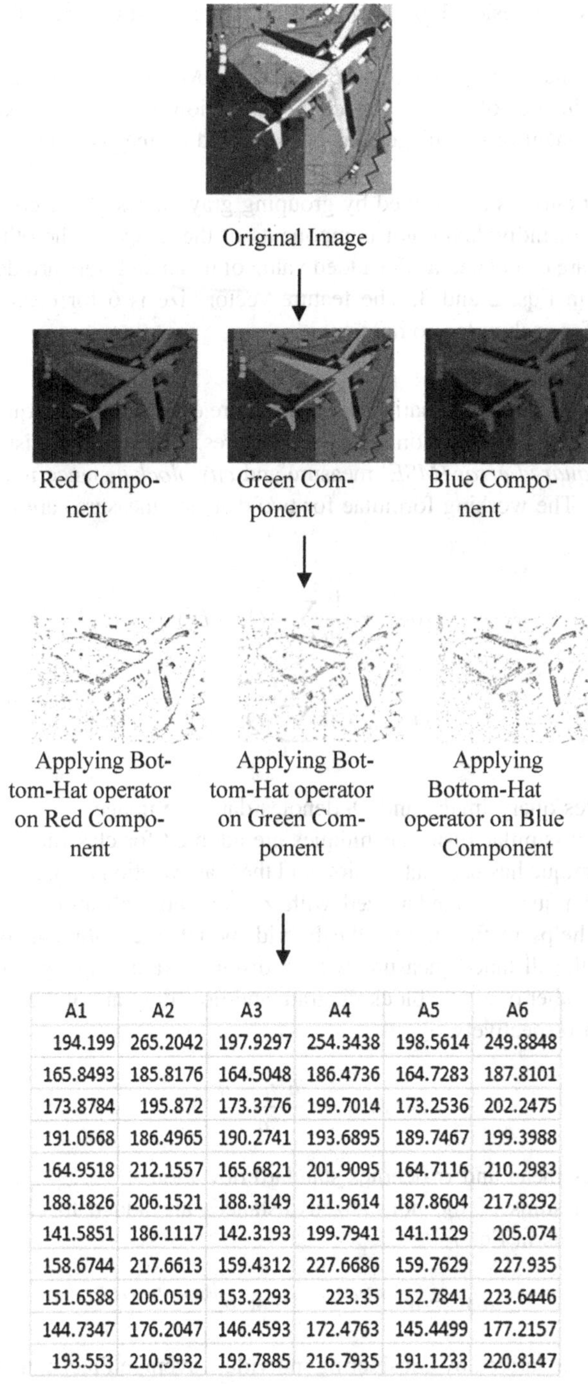

A1	A2	A3	A4	A5	A6
194.199	265.2042	197.9297	254.3438	198.5614	249.8848
165.8493	185.8176	164.5048	186.4736	164.7283	187.8101
173.8784	195.872	173.3776	199.7014	173.2536	202.2475
191.0568	186.4965	190.2741	193.6895	189.7467	199.3988
164.9518	212.1557	165.6821	201.9095	164.7116	210.2983
188.1826	206.1521	188.3149	211.9614	187.8604	217.8292
141.5851	186.1117	142.3193	199.7941	141.1129	205.074
158.6744	217.6613	159.4312	227.6686	159.7629	227.935
151.6588	206.0519	153.2293	223.35	152.7841	223.6446
144.7347	176.2047	146.4593	172.4763	145.4499	177.2157
193.553	210.5932	192.7885	216.7935	191.1233	220.8147

Extracted Feature vectors of dimension 6 per image for 11 images

Fig. 3. Process of feature extraction with *Bottom-Hat* operator

hat transform is the residual of a *closing* compared to the original signal *f*, i.e., $f - (f \bullet K)$.

The authors have implemented *Bottom-Hat* Morphological edge extraction (*BHMEE*) [38].The area of interest for feature extraction in each of the Red, Green and Blue color component of the images has been located by means of the technique as in Fig. 3.

Two different clusters are formed by grouping gray values of the contour region as one and the rest of individual color components of the image as the other group. The features vectors are calculated as the added value of mean and standard deviation of the two clusters as in Eqs. 2 and 3. The feature vector size is 6 for each image with 2 feature vectors for each color component.

Framework for Classification

At the outset, independent evaluation of each feature extraction technique is performed by means of separate classification distance measures. Two different distance measures namely, *mean squared error (MSE)* measure and *city block* distance measure are used for this purpose. The working formulae for two distance measures are in Eqs. 4 and 5 respectively.

$$D_{MSE} = \frac{1}{mn}\sum_{i=1}^{n}(Q_i - D_i)^2 \tag{4}$$

$$D_{cityblock} = \sum_{i-1}^{n}|Q_i - D_i| \tag{5}$$

where, Q_i denotes query image and D_i denotes database image.

Further, two dissimilar fusion techniques are adopted for classification.

The first technique has used late fusion and the classification decisions inferred from each of the techniques is standardized with Z score normalization as in Eq. 6. The standardization helps in fine tuning the hybrid model, preventing it to make biased decisions with the distance measure with a distance value having higher "weight" compared to the other one. The block diagram and flow diagram of the process is given in Figs. 4 and 5 respectively.

$$dist_n = \frac{dist_i - \mu}{\sigma} \tag{6}$$

where, μ denotes mean and σ denotes standard deviation

The weighted sums of the normalized distances are added to calculate the final distance measure as in Eq. 7.

$$dist = w_1 d_n^{binarization} + w_2 d_n^{morpho\,logical} \tag{7}$$

The weights are derived by considering individual average Precision rates of the two feature extraction techniques involved.

The second method has used early fusion and has standardized the feature values extracted from two different feature extraction techniques as in Figs. 6 and 7. *Z* score normalization is applied to standardize the values of feature set having concatenated feature values of the two feature extraction techniques.

Fine tuning of the classification model is carried out with the normalization process on both the fusion techniques to avoid biased classification decision for a high value feature vector which have higher influence or "weight" for the decision process. Hence, the data is normalized to a common range such as [−1, 1] or [0.0, 1.0] for uniformity. The classification is carried out using K nearest neighbor classifier (KNN), Support Vector Machine (SVM) and Random Forest (RF) classifier.

The algorithm for classification using late fusion is as follows:

1. Input an Image *I*
2. Extract features from *I* using *binarization technique/morphological operator*

   2.1 /* Extract *BINARIZATION/MORPHOLOGY feature vectors (fv)* using procedure *extract<technique>features*/*

$$fv \leftarrow extract\,BINARIZATION features(I)/extract\,MORPHOLOGICAL features(I)$$

3. Classify the binarization features using mean squared error (MSE) distance measure
4. Classify the morphological features using cityblock distance measure
5. Normalize both the distance measures using *Z* score normalization

   5.1 /* Normalized distance $dist_n$ is calculated by subtracting particular distance measure $dist_t$ from the mean μ and dividing the result by standard deviation σ

$$dist_n = \frac{dist_i - \mu}{\sigma}$$

6. Individual average Precision rates of the two feature extraction techniques are considered to calculate the weights of the distance measures.
7. Weighted sums of normalized distance measures for classification are calculated as $w_1 d_n^{binarization}$ and $w_2 d_n^{morpho\,logical}$ respectively.
8. The weighted sums are added to each other and sorted from smallest to largest to obtain final distance measure of the nearest match to the query image.

The algorithm for classification using early fusion is as follows:

1. Input an Image *I*
2. Extract features from *I* using *binarization technique/morphological operator*

   2.1 /* Extract *BINARIZATION/MORPHOLOGY feature vectors (fv)* using procedure *extract<technique>features*/*

$$fv \leftarrow extract\,BINARIZATION features(I)/extract\,MORPHOLOGICAL features(I)$$

3. Normalize both the feature vectors separately using Z score normalization

   3.1 /* Normalized feature fv_n is calculated by subtracting a particular feature value fv from the mean μ of all feature values and dividing the result by standard deviation σ*/

$$fv_n = \frac{fv - \mu}{\sigma}$$

4. The normalized feature vectors of two different techniques are concatenated horizontally to form the fused feature vector for training.
5. Classification is carried out by training the KNN/SVM/Random Forest (RF) classifiers using the fused feature vector as training data.

Framework for Retrieval

Two different retrieval techniques are designed and compared for retrieval results. The first technique has fired a generic query and has retrieved the top 20 matches from the dataset. The second technique has classified the query before forwarding it for retrieval purpose. The classified query is fired for retrieval of top 20 matches only from the identified class. On contrary to the conventional process, it has restricted the query only to the classified category to look for desired results and has limited the search space.

The steps for retrieval with generic fused query are as follows:

- Extraction of feature vectors is carried out with two different techniques, namely, feature extraction with image binarization and feature extraction with morphological operator.

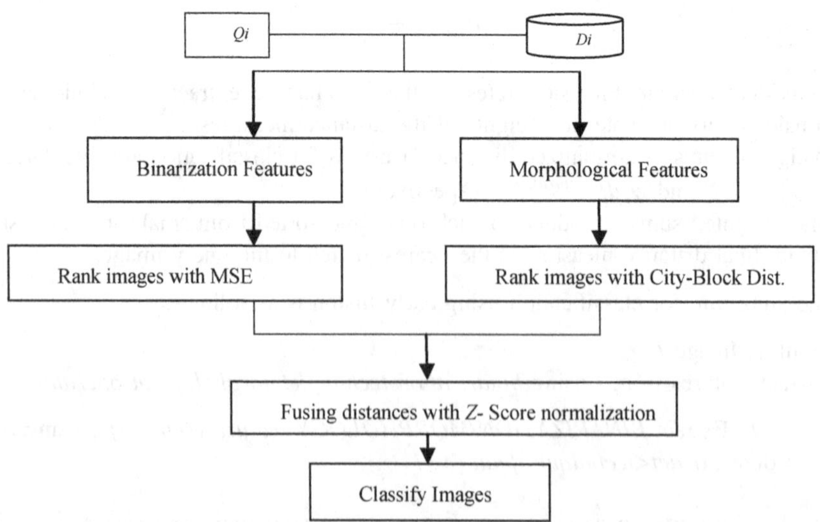

Fig. 4. Classification with late fusion framework

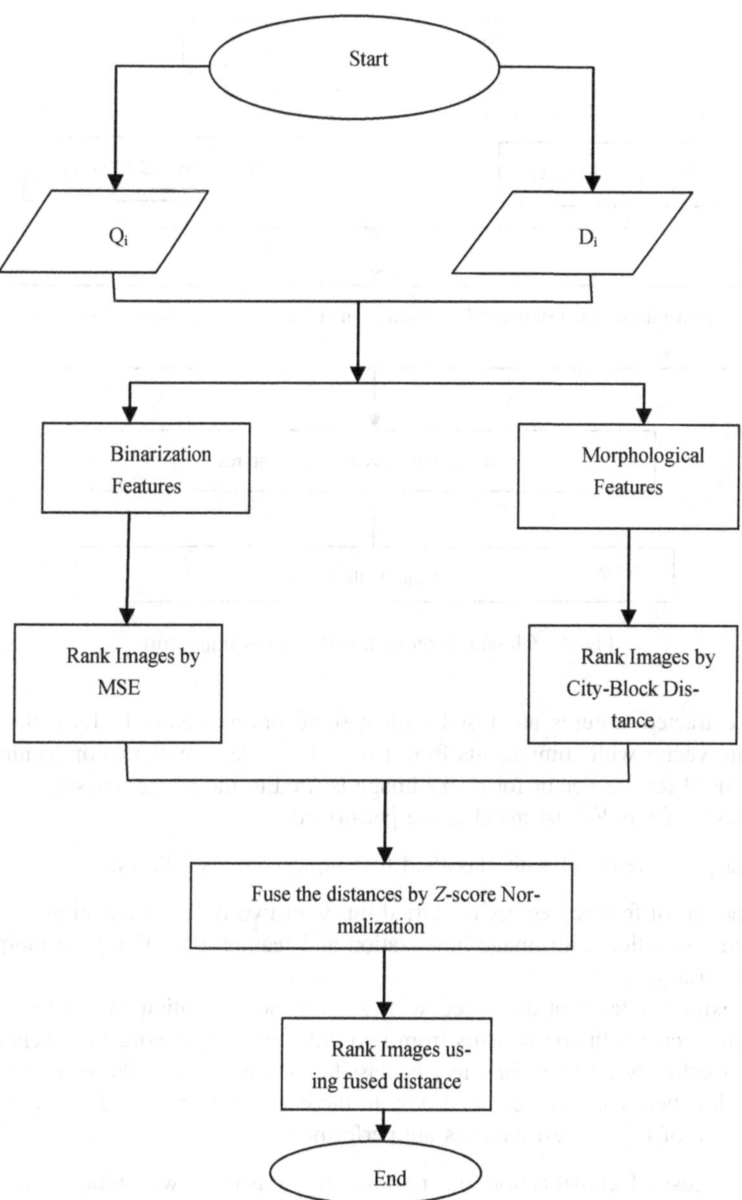

Fig. 5. Flow diagram of late fusion framework

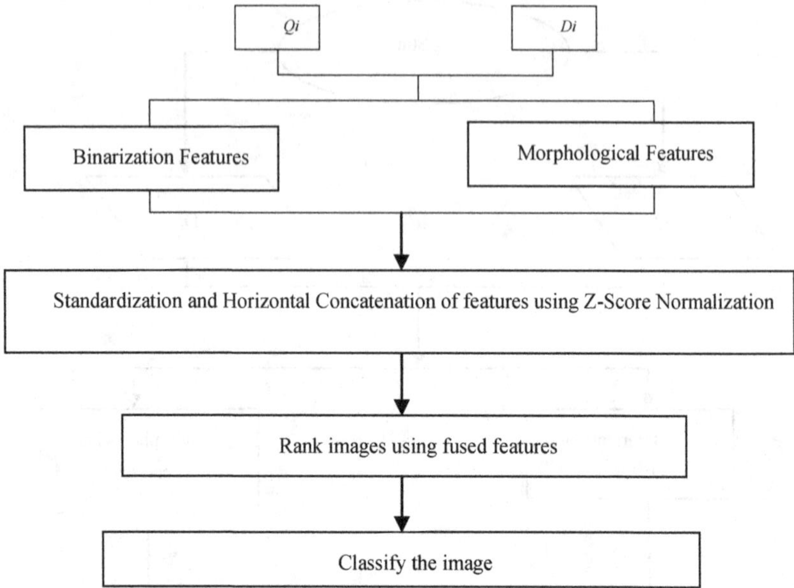

Fig. 6. Classification with early fusion framework

- The extracted features are fused with z score normalization to form the hybrid feature vector with components from two different feature extraction techniques.
- The fused feature vector for query image is fired to the image dataset.
- Retrieval of top 20 best matches are performed.

The steps for retrieval with classified fused query are as follows:

- Extraction of feature vectors is carried out with two different techniques, namely, feature extraction with image binarization and feature extraction with morphological operator.
- The extracted features are fused with z score normalization to form the hybrid feature vector with components from two different feature extraction techniques.
- The fused feature vector for query is classified to the most similar class it matches.
- The classified query is then fired only to the class of interest of the image dataset.
- Retrieval of top 20 best matches are performed.

The process of classification and retrieval are illustrated with block diagram and flow diagram in Figs. 4, 5, 6, 7 and 8.

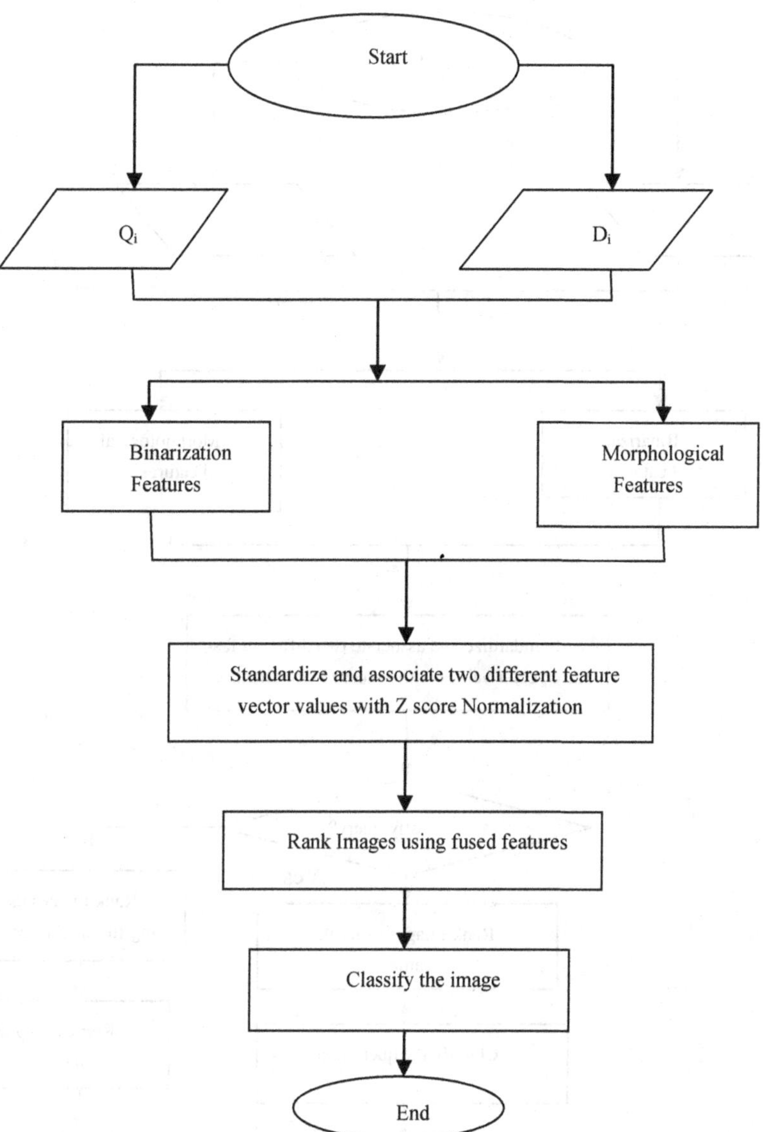

Fig. 7. Flow diagram of early fusion framework

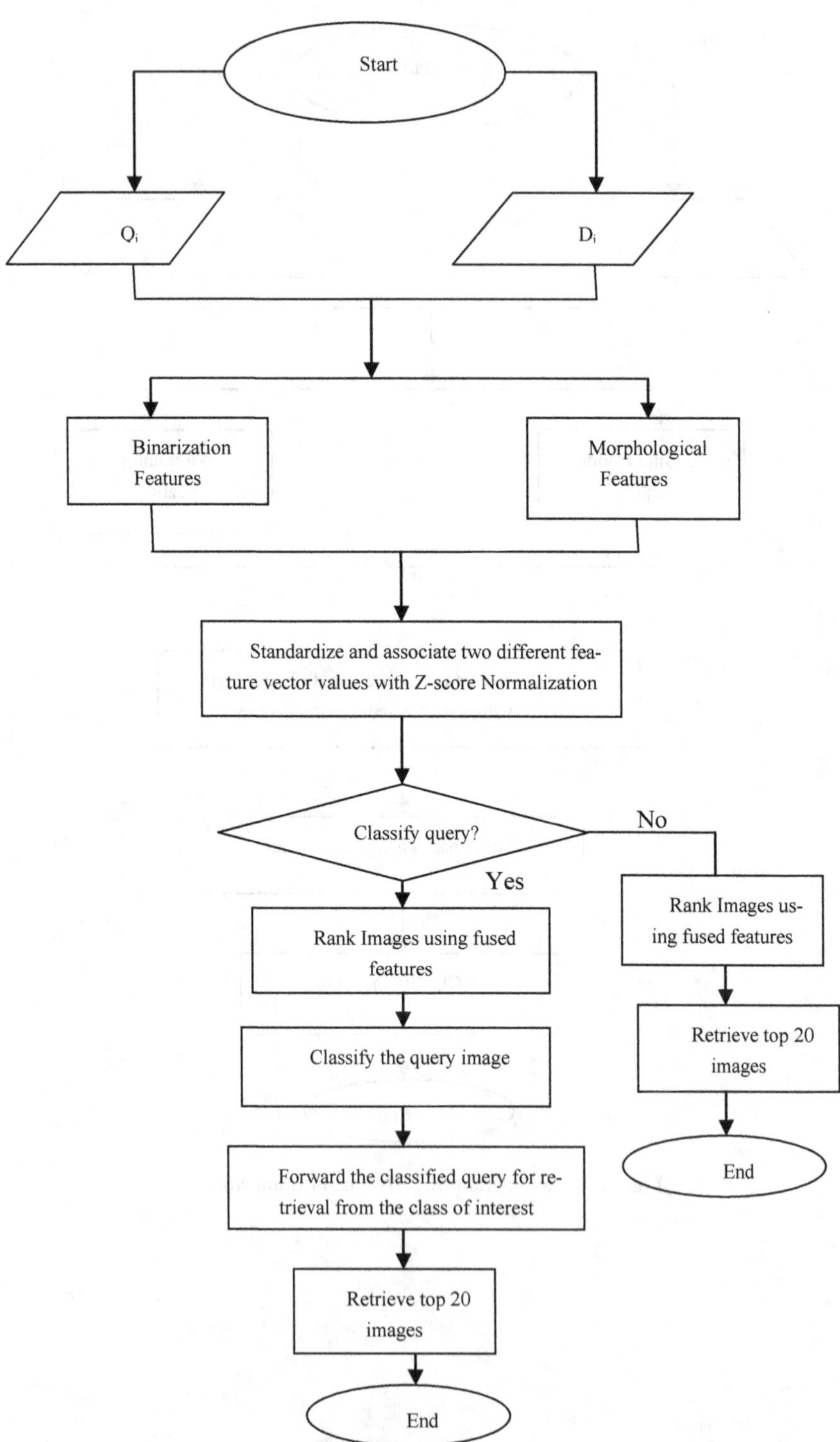

Fig. 8. Flow diagram of retrieval framework

4 Experimental Setup

Dataset

The work has considered 1000 images from a land use image dataset which is publicly available for research purpose (Yang and Shawn 2010). The images are distributed in 10 different classes having 100 images each. The 10 classes are identified as agriculture, airplane, baseballdiamond, beach, buildings, chaparral, denseresidential, forest, freeway and golfcourse. The image dimension is fixed at 256 × 256 pixels in the dataset for all images. Manual extraction is carried out for each of the images from large images in the USGS National Map Urban Area Imagery collection for various urban areas around the country. The resolution of pixel intensity is considered as 1 foot for this public domain imagery. Figure 9 has shown a sample collage of the used dataset.

Fig. 9. Sample collage of 10 categories from the USGS National Map Urban Area Imagery collection

Evaluation Techniques

Precision, Recall, F1 Score and Misclassification Rate (MR) are considered evaluation metric for classification performances. The retrieval performance isjudged by considering the mean of Precision and Recall values for different techniques of feature extraction. The metrics have followed the standard definitions and the working formulae for each of the metrics have been given in Eqs. 7, 8, 9 and 10.

$$\Pr ecision = \frac{Total..Number..of..\text{Relevant}..\text{Images}..Identified}{Total..Number..of..identified..\text{Images}} \tag{7}$$

$$\text{Recall} = \frac{Total..Number..of..\text{Relevant}..\text{Images}..Identified}{Total..Number..of..\text{Images}..in..the..\text{Relevant}..Class} \tag{8}$$

$$F1score = \frac{2 * \Pr ecision * \text{Recall}}{\Pr ecision + \text{Recall}} \tag{9}$$

$$MR = \frac{FP + FN}{TP + TN + FP + FN} \tag{10}$$

True Positive (TP) = Number of correctly classified instances.
True Negative (TN) = Number of negative results for negative instances.
False Positive (FP) = Number of positive results for negative instances.
False Negative (FN) = Number of negative results for positive instances.

5 Results and Discussions

The experimental setup comprises of Matlab (R2015b) on Intel core i5 processor with 4 GB RAM. To begin with, evaluation of classification performances is carried out individually applying two different techniques of feature extraction, namely, binarization based feature extraction technique and morphological feature extraction technique. Further, two fusion based approaches are designed and aimed to attain higher degree of successful classification by standardization of feature values and classification decision respectively of each of the participating feature extraction techniques with *Z* score normalization. The comparative categorical results for individual feature extraction techniques to that of the fusion based techniques are shown in Table 1.

It is observed in Table 1 that classification with binarization based feature extraction technique has higher Precision and Recall values compared to the feature extraction with morphological operator. Binarization techniques have locally created the threshold to separate the background and foreground of the image which has resulted in superior feature extraction compared to the feature extraction by *BottomHat* morphological operator. The category named beach has shown the highest Precision and Recall closely followed by the category called chaparral for feature extraction technique using binarization. The lowest Precision and Recall values have been revealed by the category named agricultural. Feature extraction with morphological operator has resulted in highest Precision and Recall values for the category named forest and least values for the category named baseballdiamond.

The individual techniques have shown different values for successful classification which was based on the robustness of the feature extraction techniques from the images. Further, two fusion based approaches is designed and is aimed to attain higher degree of successful classification by standardization of feature values and classification decision respectively of each of the participating feature extraction techniques with *Z* score normalization. The comparison of average Precision and Recall values for the individual techniques and the fused approach has been given in Fig. 10.

Table 1. Comparison of precision and recall for classification

	Recall (feature extraction with binarization)	Recall (feature extraction with morphological operator)	Recall (classification with late fusion)	Recall (classification with early fusion)	Precision (feature extraction with binarization)	Precision (feature extraction with morphological operator)	Precision (classification with late fusion)	Precision (classification with early fusion)	Categories
	0.39	0.6	0.76	0.77	0.500	0.759	0.835	0.846	Agricultural
	0.62	0.73	0.76	0.74	0.596	0.593	0.697	0.725	Airplane
	0.71	0.57	0.71	0.72	0.676	0.487	0.755	0.766	Baseballdiamond
	0.85	0.38	0.78	0.91	0.817	0.760	0.918	0.910	Beach
	0.65	0.67	0.75	0.74	0.596	0.798	0.735	0.763	Building
	0.96	0.95	0.98	0.99	0.814	0.664	0.916	0.952	Chaparral
	0.68	0.47	0.66	0.71	0.660	0.547	0.559	0.602	Denseresidential
	0.79	0.91	0.92	0.93	0.775	0.843	0.860	0.877	Forest
	0.44	0.58	0.48	0.51	0.698	0.547	0.828	0.785	Freeway
	0.69	0.6	0.79	0.83	0.605	0.577	0.612	0.675	Golfcourse
Average	0.678	0.646	0.759	0.785	0.674	0.658	0.772	0.79	

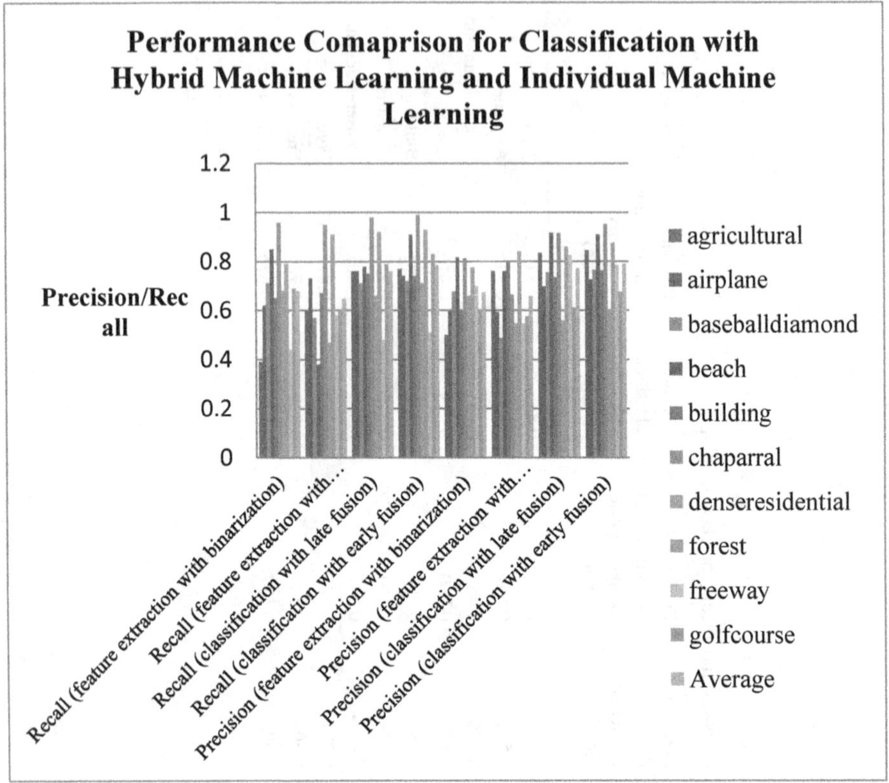

Fig. 10. Comparison of precision and recall values for individual techniques and fused approach

It is observed in Fig. 10 that both the fusion based machine learning techniques have outclassed the individual ones and have revealed enhanced Precision and Recall values. Individual feature extraction techniques are not able to capture important information contributing towards enhanced classification results compared to the fusion based approach. The fused approach has combined information gathered by two different feature extraction techniques which has facilitated in creation of a final robust feature vector for training purpose of the machine learning model.

Further, comparing the Precision and Recall values of the two fusion based approaches, namely the late fusion approach and the early fusion approach, it is revealed that the Precision and Recall values with the early fusion approach has outperformed the Precision and Recall values of the late fusion technique.

The feature vector created with early fusion technique is evaluated with three different classifiers, namely, KNN, SVM and Random Forest (RF). Comparative results are shown in Fig. 11.

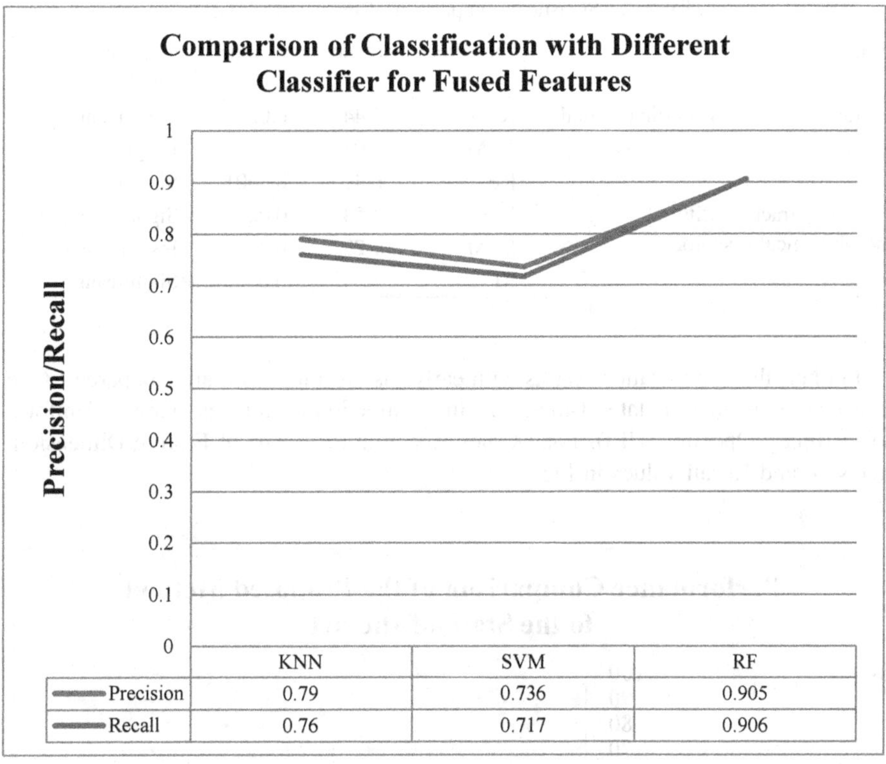

Fig. 11. Comparison of precision and recall values with different classifiers for fused features

It is observed, that, Random Forest (RF) classifier has portrayed the highest classification results followed by KNN classifier. SVM has the least results.

Henceforth, a statistical test is conducted to evaluate the significance of the fusion based approach over individual approaches. A hypothesis testing method is adopted.

Hypothesis

Precision values of classification for the individual techniques do not differ to that of the precision value of classification of the fused technique.

A paired *t*-test is carried out and the results are shown in Table 2. The *p*-values calculated for classification with KNN and Random Forest (RF) classifier by training them with fused feature have shown significance over individual techniques. Therefore, the null hypothesis is rejected and statistical significance of improvements in classification using early fusion is established. However, classification with SVM is not found to be significant.

Table 2. Statistical comparison of precision values

Features	Classifiers	t-calc	p-value	Statistical significance
Feature extraction with binarization	KNN	2.44	0.02	Significant
	SVM	1.04	0.31	Insignificant
	RF	6.46	0.0001	Significant
Feature extraction with morphological operator	KNN	2.54	0.02	Significant
	SVM	1.23	0.23	Insignificant
	RF	5.94	0.0001	Significant

Further, the classification results with early fusion framework are compared to the classification results of state-of-the-art feature extraction techniques, namely, Uniform Local Binary Operator (ULBP) and Color Histogram in terms of Feature Dimension, Precision and Recall values in Fig. 12.

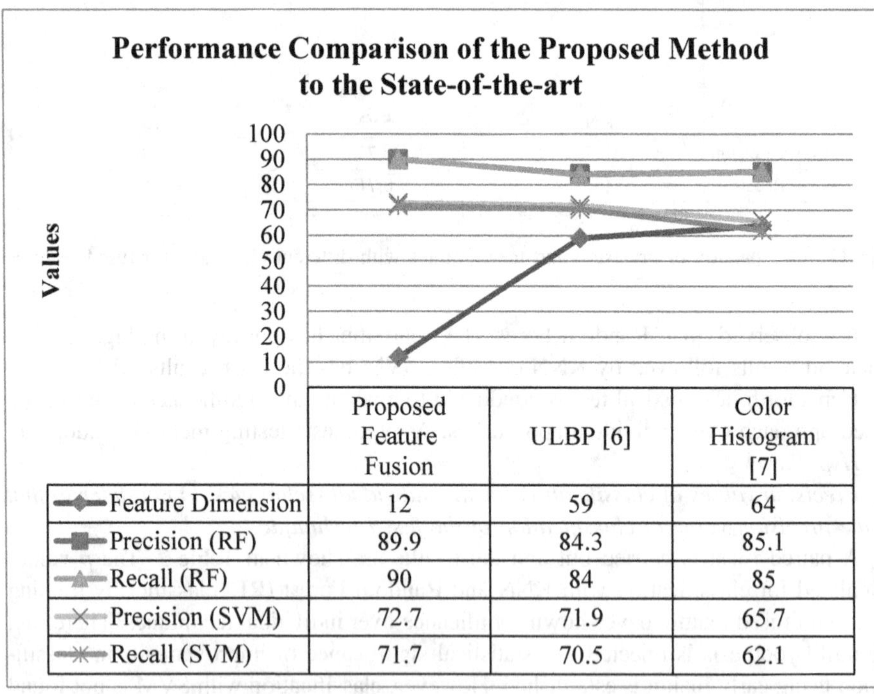

Fig. 12. Comparison of feature dimension, precision and recall of proposed technique with existing approaches for content based remotely sensed image classification

Table 3. Comparison of precision values for retrieval with individual techniques, multi technique fusion with generic query and multi technique fusion with classified query for wang dataset

Categories	Precision (feature extraction with binarization)	Recall (feature extraction with binarization)	Precision (feature extraction with morphological operator)	Recall (feature extraction with morphological operator)	Precision (early fusion using generic query)	Recall (early fusion using generic query)	Precision (early fusion using classified query)	Recall (early fusion using classified query)
Agricultural	35	7	61	12.2	63	12.6	80	16
Airplane	45	9	58	11.6	61	12.2	100	20
Baseballdiamond	53	10.6	54	10.8	66	13.2	100	20
Beach	91	18.2	32	6.4	98	19.6	100	20
Building	46	9.2	40	8	59	11.8	80	16
Chaparral	76	15.2	77	15.4	90	18	80	16
Denseresidential	50	10	37	7.4	65	13	80	16
Forest	84	16.8	99	19.8	97	19.4	100	20
Freeway	60	12	42	8.4	68	13.6	80	16
Golfcourse	37	7.4	40	8	57	11.4	60	12
average	57.7	11.54	54	10.8	72.4	14.48	86	17.2

The results in Fig. 12 has shown that the feature dimension of the proposed technique is 12, whereas, the existing techniques, namely, ULBP and Color Histogram have feature dimension of 59 and 64 respectively. Moreover, it is also observed that the precision and recall values for classification with the proposed feature fusion technique have outclassed the results for ULBP and CH. Two different classifiers, namely, Random Forest (RF) and Support Vector Machine (SVM) are used for the assessment of classification results. The proposed technique has outperformed the existing ones in both the cases.

Henceforth, the feature extraction techniques are tested for the retrieval performances. The retrieval is carried out in two different ways, namely, by generic query and by classified query. Random selection of 5 different images is performed from each category of the dataset. On the whole, 50 images were selected as query images from all the 10 categories of the dataset. Firstly, a generic query is fired at the dataset and top 20 matches were selected for two different feature extraction techniques individually. Henceforth, an early fusion is performed with the feature values of two different techniques based on Z score normalization and early fusion based generic query is fired at the dataset to retrieve top 20 matches. Finally, the early fusion based query is classified to its nearest matching class and is forwarded to search for results only within the class of interest.

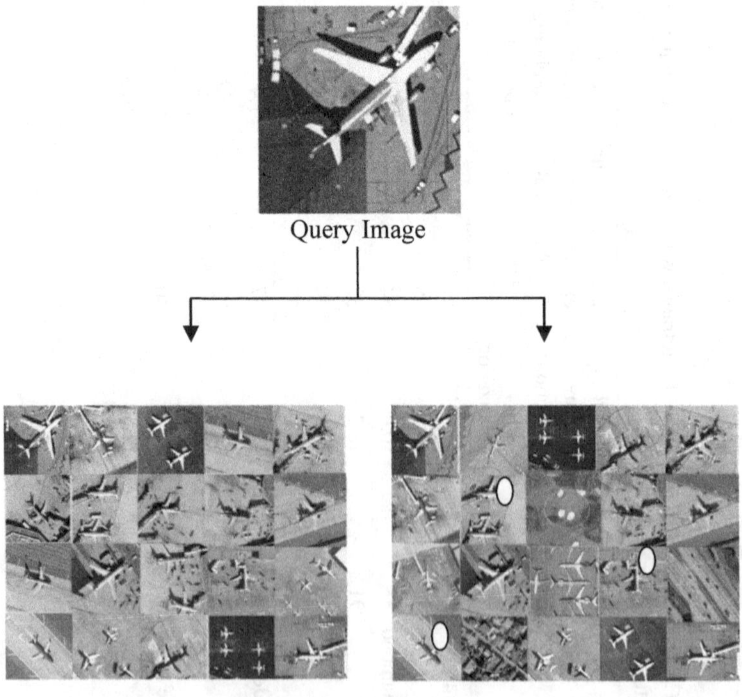

Fig. 13. Comparison of retrieval with classified query and generic query

A comparison of category wise Precision and Recall values for retrieval with generic query and classified query is shown in Table 3. Sample retrieval is shown in Fig. 12 which has compared retrieval with generic query with respect to classified query.

The results in Table 3 have clearly revealed that retrieval with fusion framework has higher Precision and Recall values compared to individual techniques. Further, it is also observed that retrieval with early fusion using classified query has yielded higher Precision and Recall values compared to retrieval with early fusion using generic query. The Precision for retrieval by early fusion using classified query is 13.6% higher compared to that of early fusion with generic query as in Table 3.

Figure 13 has evidently exposed that retrieval with classified query has identified all the 20 images from the same category of the query image, whereas, retrieval with generic query has recognized 17 images from the category of query image and 3 different images from the category named forest, freeway and dense-residential which is marked with white ellipses.

6 Conclusion

The paper has carried out exhaustive comparison of different feature extraction techniques and the usefulness of the techniques to predict classification and retrieval results. Consequently, the following conclusions are enlisted:

- Successful proposition of two different feature extraction techniques based on image binarization and morphological operators.
- The dimension of feature vectors is radically reduced for each feature extraction technique.
- The feature size is independent of the magnitude of the image.
- Designing the fusion framework of machine learning for both classification and retrieval methods

The importance of content based image classification using advanced machine learning techniques is evident in this work. It has also highlighted the expediency of classification as a precursor for query categorization for retrieval. It is inferred that early fusion framework with multi feature extraction techniques has enhanced capabilities for content based information identification with remotely sensed image data with comparatively lesser computational overhead. Nevertheless, we are aiming to design deep learning based approaches for feature extraction from diverse image categories to further enhance classification accuracy.

References

1. Maxwell, A.E., Warner, T.A., Fang, F.: Implementation of machine-learning classification in remote sensing: an applied review. Int. J. Remote Sens. **39**(9), 2784–2817 (2018)
2. Cai, Y., et al.: A high-performance and in-season classification system of field-level crop types using time-series Landsat data and a machine learning approach. Remote Sens. Environ. **210**, 35–47 (2018)
3. Yang, J., Wong, M.S., Ho, H.C.: Retrieval of urban surface temperature using remote sensing satellite imagery. In: Dey, N., Bhatt, C., Ashour, Amira S. (eds.) Big Data for Remote Sensing: Visualization, Analysis and Interpretation, pp. 129–154. Springer, Cham (2019). https://doi.org/10.1007/978-3-319-89923-7_5
4. Das, R., Walia, E.: Partition selection with sparse autoencoders for content based image classification. Neural Comput. Appl. **31**, 675–690 (2017)
5. Zhao, W., Du, S.: Spectral–spatial feature extraction for hyperspectral image classification: a dimension reduction and deep learning approach. IEEE Trans. Geosci. Remote Sens. **54**(8), 4544–4554 (2016)
6. Li, Y., Zhang, Y., Tao, C., Zhu, H.: Content-based high-resolution remote sensing image retrieval via unsupervised feature learning and collaborative affinity metric fusion. Remote Sens. **8**(9), 709 (2016)
7. Zhang, Y., Yang, X., Cattani, C., Rao, R.V., Wang, S., Phillips, P.: Tea category identification using a novel fractional Fourier entropy and Jaya algorithm. Entropy **18**(3), 77 (2016)
8. Gonzales-Barron, U., Butler, F.: A comparison of seven thresholding techniques with the k-means clustering algorithm for measurement of bread-crumb features by digital image analysis. J. Food Eng. **74**(2), 268–278 (2006)
9. Li, H., Liu, L., Huang, W., Yue, C.: An improved fusion algorithm for infrared and visible images based on multi-scale transform. Infrared Phys. Technol. **74**, 28–37 (2016)
10. Kumar, S., Toshniwal, D.: Analysis of hourly road accident counts using hierarchical clustering and cophenetic correlation coefficient (CPCC). J. Big Data **3**(1), 13 (2016)
11. Zhang, L., Li, A., Zhang, Z., Yang, K.: Global and local saliency analysis for the extraction of residential areas in high-spatial-resolution remote sensing image. IEEE Trans. Geosci. Remote Sens. **54**(7), 3750–3763 (2016)
12. Tang, J., Woods, M., Cossell, S., Liu, S., Whitty, M.: Non-productive vine canopy estimation through proximal and remote sensing. IFAC- Papers On-Line **49**(16), 398–403 (2016)
13. Valizadeh, M., Armanfard, N., Komeili, M., Kabir, E.: A novel hybrid algorithm for binarization of badly illuminated document images. In: 2009 14th International CSI Computer Conference, CSICC 2009, pp. 121–126. IEEE, October 2009
14. Pitkänen, J.: Individual tree detection in digital aerial images by combining locally adaptive binarization and local maxima methods. Can. J. For. Res. **31**(5), 832–844 (2001)
15. Liu, H., Jezek, K.C.: Automated extraction of coastline from satellite imagery by integrating Canny edge detection and locally adaptive thresholding methods. Int. J. Remote Sens. **25**(5), 937–958 (2004)
16. Al-Amri, S.S., Kalyankar, N.V.: Image segmentation by using threshold techniques. arXiv preprint arXiv:1005.4020 (2010)
17. Rosin, P.L., Ioannidis, E.: Evaluation of global image thresholding for change detection. Pattern Recogn. Lett. **24**(14), 2345–2356 (2003)

18. Manno-Kovács, A., Ok, A.O.: Building detection from monocular VHR images by integrated urban area knowledge. IEEE Geosci. Remote Sens. Lett. **12**(10), 2140–2144 (2015)
19. Liu, H., He, G.: Shape feature extraction of high resolution remote sensing image based on susan and moment invariant. In: Processing of the 2008 Congress on Image and Signal, CISP 2008, vol. 2, pp. 801–807. IEEE, May 2008
20. Ezer, T., Liu, H.: On the dynamics and morphology of extensive tidal mudflats: integrating remote sensing data with an inundation model of Cook Inlet, Alaska. Ocean Dyn. **60**(5), 1307–1318 (2010)
21. Neubert, M., Herold, H., Meinel, G.: Evaluation of remote sensing image segmentation quality–further results and concepts. Int. Arch. Photogramm. Remote Sens. Spat. Inf. Sci. **36** (4/C42) (2006)
22. Katartzis, A., Sahli, H.: A stochastic framework for the identification of building rooftops using a single remote sensing image. IEEE Trans. Geosci. Remote Sens. **46**(1), 259–271 (2008)
23. Eismann, M.T.: Hyperspectral Remote Sensing. SPIE, Bellingham (2012)
24. Elmahdy, S.I., Mansor, S., Huat, B.B., Mahmod, A.R.: Structural geologic control with the limestone bedrock associated with piling problems using remote sensing and GIS: a modified geomorphological method. Environ. Earth Sci. **66**(8), 2185–2195 (2012)
25. Cipolletti, M.P., Delrieux, C.A., Perillo, G.M., Piccolo, M.C.: Superresolution border segmentation and measurement in remote sensing images. Comput. Geosci. **40**, 87–96 (2012)
26. Forestier, G., Puissant, A., Wemmert, C., Gançarski, P.: Knowledge-based region labeling for remote sensing image interpretation. Comput. Environ. Urban Syst. **36**(5), 470–480 (2012)
27. Hong, G., Zhang, Y., Mercer, B.: A wavelet and IHS integration method to fuse high resolution SAR with moderate resolution multispectral images. Photogramm. Eng. Remote Sens. **75**(10), 1213–1223 (2009)
28. Zhou, X., Liu, J., Liu, S., Cao, L., Zhou, Q., Huang, H.: A GIHS-based spectral preservation fusion method for remote sensing images using edge restored spectral modulation. J. Photogramm. Remote Sens. **88**, 16–27 (2014)
29. Ling, Y., Ehlers, M., Usery, E.L., Madden, M.: FFT-enhanced IHS transform method for fusing high-resolution satellite images. ISPRS J. Photogramm. Remote Sens. **61**(6), 381–392 (2007)
30. Zhang, L., Li, Y., Lu, H., Yamawaki, A., Yang, S., Serikawa, S.: Maximum local energy method and sum modified Laplacian for remote image fusion based on beyond wavelet transform. Appl. Math. Inf. Sci. **7**(1S), 149–156 (2013)
31. Zhu, Q., Shyu, M.L.: Sparse linear integration of content and context modalities for semantic concept retrieval. IEEE Trans. Emerg. Top. Comput. **3**(2), 152–160 (2015)
32. Byun, Y., Choi, J., Han, Y.: An area-based image fusion scheme for the integration of SAR and optical satellite imagery. IEEE J. Sel. Top. Appl. Earth Obs. Remote Sens. **6**(5), 2212–2220 (2013)
33. Su, Y., Lee, C.H., Tu, T.M.: A multi-optional adjustable IHS-BT approach for high resolution optical and SAR image fusion. Chung Cheng Ling Hsueh Pao/J. Chung Cheng Inst. Technol. **42**(1), 119–128 (2013)
34. Choi, J., Yeom, J., Chang, A., Byun, Y., Kim, Y.: Hybrid pansharpening algorithm for high spatial resolution satellite imagery to improve spatial quality. IEEE Geosci. Remote Sens. Lett. **10**(3), 490–494 (2013)
35. Pohl, C., van Genderen, J.: Remote sensing image fusion: an update in the context of digital earth. Int. J. Digit. Earth **7**(2), 158–172 (2014)

36. Rokni, K., Ahmad, A., Solaimani, K., Hazini, S.: A new approach for surface water change detection: integration of pixel level image fusion and image classification techniques. Int. J. Appl. Earth Obs. Geoinf. **34**, 226–234 (2015)
37. Feng, M.L., Tan, Y.P.: Adaptive binarization method for document image analysis. In: 2004 IEEE International Conference on Multimedia and Expo, ICME 2004, vol. 1, pp. 339–342. IEEE, June 2004
38. Jalba, A.C., Wilkinson, M.H., Roerdink, J.B.: Morphological hat-transform scale spaces and their use in pattern classification. Pattern Recogn. **37**(5), 901–915 (2004)

Clustering-Based Aggregation of High-Utility Patterns from Unknown Multi-database

Abhinav Muley(✉) and Manish Gudadhe

St. Vincent Pallotti College of Engineering and Technology, Nagpur, India
abhi.muley27@gmail.com, mbgpatil@gmail.com

Abstract. High-utility patterns generated from mining the unknown and different databases can be clustered to identify the most valid patterns. Sources include the internet, journals, and enterprise data. Here, a grid-based clustering method (CLIQUE) is used to aggregate patterns mined from multiple databases. The proposed model forms the clusters based on all the utilities of patterns to determine the interestingness and the correct interval of its utility measure. The set of all patterns is collected by first mining the databases individually, at the local level. The problem arises when the same pattern is identified by all of the databases but with different utility factors. In this case, it becomes difficult to decide whether the pattern should be considered as a valid or not, due to the presence of multiple utility values. Hence, an aggregation model is applied to test whether a pattern satisfies the utility threshold set by a domain expert. We found that the proposed aggregation model effectively clusters all of the interesting patterns by discarding those patterns that do not satisfy the threshold condition. The proposed model accurately optimizes the utility interval of the valid patterns.

Keywords: Data mining · Clustering · CLIQUE · Aggregation · High-utility patterns · Multiple databases · Distributed data mining · Multi-database mining

1 Introduction

Data mining also referred to as Knowledge Discovery in Databases (KDD), is the discovery of interesting, valid, meaningful and previously unknown patterns from a large amount of data. The implication of interestingness depends upon the application domain. Big companies can take advantage of data mining technology having great potential to help them to predict behavior and trends among consumers.

Data mining helps businesses to make proactive decisions by extracting predictive information. Enterprises can rip the benefits of data mining techniques by employing parallel processing systems to mine the vast databases of big enterprises having multiple chains [1]. The decision-making quality of businesses has been greatly influenced by a large amount of data available on the World Wide Web [2]. The corporates are also taking the benefit of an intranet to collect and share the data inside the corporation. Post-processing of mined data is also necessary to analyze and maintain the useful knowledge discovered. Cluster analysis and pattern analysis helps to analyze the

M. L. Gavrilova and C. J. K. Tan (Eds.): Trans. on Comput. Sci. XXXIV, LNCS 11820, pp. 29–43, 2019.
https://doi.org/10.1007/978-3-662-59958-7_2

changes occurring in the patterns mined from different databases. Cluster analysis or clustering can be employed to aggregate the different local patterns discovered from multiple databases for global pattern analysis. However, a large number of patterns obtained at the global level can become a bottleneck to identify truly useful patterns. This problem can be solved by performing a cluster analysis of the patterns based on its utility measure.

Knowledge discovery in multiple databases is recognized as an important research topic in data mining community [3–7]. Existing research has focused on developing parallel and distributed data mining algorithms for applications with multiple databases. Most of the parallel data mining algorithms in the literature focus on mining the rules or patterns from known data sources. One of the examples is synthesizing frequent rules by allocating weights to data sources based on Good's idea [15], proposed by Wu et al. [4]. In the existing work, the issue of unknown database mining has not been addressed. In real life applications, the sources can forward their retrieved patterns at the central node but may not reveal their identity or locations. If the sources of retrieved patterns are unknown, the weights of patterns and data sources can't be determined to synthesize the patterns. In this paper, we have proposed an aggregation model for integrating high-utility itemset (HUIM) mined from unknown databases/sources of the similar organization that may not want to disclose their data or identity. Our proposed model effectively optimizes the utility interval for the novel/useful patterns by using a grid-based clustering method of CLIQUE (CLustering by QUEst). This has been clearly established by extensively experimenting on different benchmark datasets.

The paper is organized as follows: Sect. 1 introduces to the concepts of data mining, multiple-database mining with known and unknown sources. Section 2 describes the concepts of HUIM and clustering techniques. In Sect. 3, the problem formulation is defined. Section 4 describes the proposed aggregation model with an example of the real dataset. Section 5 describes the effectiveness of the experiments performed.

2 Related Work

The association rule mining based parallel and distributed techniques [16–21, 26–28] have several drawbacks as they consider only frequent itemsets for mining and doesn't perform local pattern analysis [4]. The Frequent Itemset Mining (FIM) [22] discovers the itemsets appearing most frequently in the transaction database. It has two fundamental limitations as follows: (1) FIM assumes that every item in the transaction has the same utility in terms of profit/item. (2) FIM assumes that every item appears only once in any transaction. This is contrary to many applications where an item appears in multiple quantities. Hence, FIM is unable to capture the novel patterns in the databases where an item appears multiple times in a single transaction. The solution to this problem is addressed in high-utility itemset mining (HUIM) [23–25]. The support-confidence measure in FIM is replaced by the utility (denoted as *util*) in HUIM. The most efficient algorithm for capturing high-utility itemset found in our literature survey is EFIM algorithm [14]. This algorithm is used in our experiments for performing local pattern analysis. Once the local patterns are captured, the aggregation model is performed using clustering.

Clustering is the process of dividing objects into homogeneous groups or clusters based on their attribute values in such a way that the objects in the same cluster are similar to each other, but are dissimilar to the objects from the other clusters. Cluster analysis or clustering is an unsupervised learning method which is also called as the automatic classification because the information about the class label is missing. Clustering methods are widely studied under statistics [8] and machine learning [9, 10]. It has many applications in the field of Business Intelligence (BI), web search, pattern recognition and security [11]. Clustering methods are classified into two categories: partition based and hierarchical based. However, these techniques do not address all the requirements of clustering techniques. Hence, a grid-based subspace clustering technique called as CLIQUE (CLustering by QUEst) is developed by Agrawal et. al. [12]. The grid-based method uses a multidimensional grid data structure. This results in faster processing time which is dependent on the number of cells in each dimension rather than the number of data objects. However, it is significant to perform cluster analysis within the subspaces of the data. CLIQUE is used for finding dense clusters in subspaces. It partitions data objects into cells by creating non-overlapping intervals for each dimension. Dense cells and sparse cells are separated by a density threshold [11]. A cell is considered as a dense cell if it exceeds the density threshold. Clustering is performed in CLIQUE by following steps:

- Find subspaces which contain clusters
- Identify dense clusters
- Generate minimal description using the Minimum Description Length (MDL) principle.

3 Problem Formulation

The problem for this research is defined as follows: *For any pattern PQ, the collected utilities are* $util_1, util_2, util_3, \ldots, util_n$ *from 'n' multiple databases/sources after the pattern mining is performed at individual source (local pattern analysis). We have to cluster the collected patterns to find the novel patterns satisfying the utility threshold and obtain the optimized normal distribution interval.*

We can roughly aggregate these utilities by taking an average of these, but the average operator gives the mid value among lowest and highest utilities which is a rough approach and is not necessarily always a true measure. Figure 1 illustrates our aggregation model when the high-utility patterns are discovered from different data sources of the same organization.

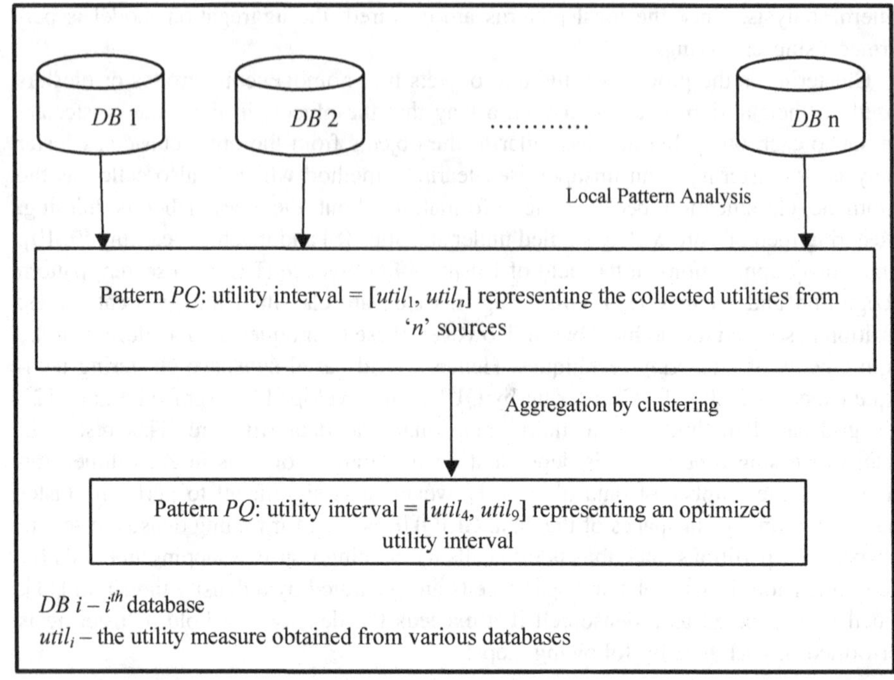

Fig. 1. The aggregation model

4 Aggregating Patterns by Clustering

An Example: From '*n*' unknown data sources/databases, suppose we have the collected pattern *PQ* with utilities $util_1 = 202{,}029$, $util_2 = 182{,}788$, $util_3 = 158{,}737$, $util_4 = 129{,}876$, and $util_5 = 288{,}613$. We are interested to find if the pattern *PQ* is a novel pattern by checking if it satisfies the minimum utility threshold. And, also optimize the utility interval by using grid-based clustering. This particular example is taken out from a retail dataset (see Sect. 5.1) with actual values. Table 1 describes all the symbols and notations used in the paper.

Table 1. The list of symbols and notations used in this paper.

Symbol	Description
n	Number of data sources
$util_i$	Utility measure associated with every pattern collected from various unknown sources
Maximum utility	The maximum value of utility in the transaction database
$nutil_i$	Normalized utility = $util_i$/*Maximum utility*
x, y	Boundary values in utility interval
K	Number of utilities belonging to the interval $[x, y]$

(*continued*)

Table 1. (*continued*)

Symbol	Description
μ	The threshold for a voting degree e.g. $\mu = 0.3$ implies patterns collected by at least 30% of sources
$P\{x \leq X \leq y\}$	Probability function of all 'X' belonging to the interval $[x, y]$
X	Random variable of probability P
β	The threshold for optimization degree
minutil	Minimum utility threshold
normalized minutil	Normalized minimum utility threshold = *minutil/Maximum utility*
$AV_{i,\,j}$	Adjacency value between $nutil_i$ and $nutil_j$
$AM(nutil_i, nutil_j)$	Adjacency measure between $nutil_i$ and $nutil_j$

4.1 Normalizing the Utilities

According to the Good's idea of the weight of evidence [15], we consider the utilities in the range [0, 1] for simplicity. The given utilities are normalized with respect to the maximum utility in that transaction database. Normalizing is done for obtaining utilities in the normal distribution interval. The maximum utility in the example database denoted by *Maximum utility* = 481,021. The normalized utilities are $nutil_1 = 0.42$, $nutil_2 = 0.38$, $nutil_3 = 0.33$, $nutil_4 = 0.27$, and $nutil_5 = 0.6$; where $nutil_i = util_i/Maximum\ utility$. These utilities are treated in increasing order of their arithmetic value. The current interval of the utility of pattern *PQ* is [0.27, 0.6].

4.2 Calculate the Adjacency Values Among Utilities

Let k be the number of utilities belonging to an interval $[x, y]$, where $0 \leq x \leq y \leq 1$. For n databases, if $k/n \geq \mu$, then these utilities are considered to be in a normal distribution where $0 < \mu \leq 1$ and is a threshold set by an expert. This implies that the interval $[x, y]$ is taken as the utility of pattern *PQ*. Suppose we need a probability $P\{x \leq X \leq y\}$, where X is a random variable valued from $nutil_1, nutil_2, nutil_3, \ldots, nutil_n$, to satisfy $P \geq \mu$ and $|y - x| \leq \beta$, where β is a user threshold. Let $\mu = 0.3$, $\beta = 0.1$, *minutil* = 144,307 (minimum utility threshold), *normalized minutil* = 0.3 (*minutil/Maximum utility* in database) for the running example. The adjacency value between $nutil_i$ and $nutil_j$ be denoted as $AV_{i,j}$ and given by

$$AV_{i,j} = 1 - \left|nutil_i - nutil_j\right| \tag{1}$$

The adjacency value among any two utilities can be calculated by representing all the utilities in a subspace 2-dimensional matrix of $n \times n$ dimensions as shown in Table 2.

Table 2. Adjacency value grid

	$nutil_1$	$nutil_2$	...	$nutil_n$
$nutil_1$	$AV_{1,1}$	$AV_{1,2}$	...	$AV_{1,n}$
$nutil_2$	$AV_{2,1}$	$AV_{2,2}$	...	$AV_{2,n}$
⋮	⋮	⋮	⋮	⋮
$nutil_n$	$AV_{n,1}$	$AV_{n,2}$	...	$AV_{n,n}$

The adjacency value is calculated as per Eq. (1) for 5 different utilities and is placed in the adjacency value grid. For example, $AV_{1,2} = 1 - |nutil_1 - nutil_2| = 1 - |0.42 - 0.38| = 0.96$. For a normal distribution, we consider the normalized utility in all the future calculations. All the results can be placed in a similar resultant matrix of $n \times n$ dimensions which is symmetric in nature as shown in Table 3.

Table 3. Adjacency value grid

	$nutil_1$	$nutil_2$	$nutil_3$	$nutil_4$	$nutil_5$
$nutil_1$	1	0.96	0.91	0.85	0.82
$nutil_2$	0.96	1	0.95	0.89	0.78
$nutil_3$	0.91	0.95	1	0.94	0.73
$nutil_4$	0.85	0.89	0.94	1	0.67
$nutil_5$	0.82	0.78	0.73	0.67	1

4.3 Calculate the Adjacency Measure Among Utilities

We use the grid-based clustering method to obtain the utility interval for a pattern. To calculate the adjacency between utilities, an adjacency measure is required. The measure determines the adjacency degree between any two utilities by adjacency values. The simple adjacency measure is defined as:

$$AM(nutil_i, nutil_j) = \sum (AV_{m,i} * AV_{m,j}) \qquad (2)$$

where 'm' takes the summation over the set of all the utilities. The results are stored in the $n \times n$ utility grid which is symmetric.

The adjacency measure is calculated as per Eq. (2) for 5 different utilities and is placed in the adjacency measure grid given in Table 4. The diagonal values are kept empty since it represents the auto-relation of utility to itself. For example, $AM(nutil_1, nutil_2) = \sum(AV_{m,1} * AV_{m,2})$ (for $m = 1, 2, \ldots, 5$) $= (1 * 0.96) + (0.96 * 1) + (0.91 * 0.95) + (0.85 * 0.89) + (0.82 * 0.78) = 4.18$.

Table 4. Adjacency measure grid

	$nutil_1$	$nutil_2$	$nutil_3$	$nutil_4$	$nutil_5$
$nutil_1$		4.18	4.13	3.96	3.623
$nutil_2$	4.18		4.18	4.01	3.64
$nutil_3$	4.13	4.18		3.99	3.58
$nutil_4$	3.96	4.01	3.99		3.42
$nutil_5$	3.623	3.64	3.58	3.42	

4.4 Calculate the Utility-Adjacency Relationship

We assume threshold to be 4.0 which identifies if two utilities are adjacent enough to be in the same cluster of each other. A binary grid of utility-adjacency relation is created to identify the dense clusters according to the CLIQUE technique. This binary grid is shown in Table 5. The binary value is 1 if $AM(util_i, util_j)$ > threshold or 0 if $AM(util_i, util_j)$ < threshold.

Table 5. Adjacency binary grid

	$nutil_1$	$nutil_2$	$nutil_3$	$nutil_4$	$nutil_5$
$nutil_1$		1	1	0	0
$nutil_2$	1		1	1	0
$nutil_3$	1	1		0	0
$nutil_4$	0	1	0		0
$nutil_5$	0	0	0	0	

Dense cluster

4.5 Forming the Clusters and Finding Out the Utility Interval

We adopted a CLIQUE technique to form the dense clusters from binary grid constructed by normally distributing the utilities. All the utilities in a cluster must be in the threshold of all other utilities. Applying the cluster algorithm to our example, we create the following clusters:

Cluster 1	$nutil_1$, $nutil_2$ and $nutil_3$
Cluster 2	$nutil_4$
Cluster 3	$nutil_5$

- For Cluster 1: $x = 0.33$ (lowest utility of cluster 1), $y = 0.42$ (highest utility of cluster 1)

$|y - x| = |0.42 - 0.33| = 0.09 < \beta(0.1)$, $P\{x \le X \le y\} = 3/5 = 0.6 > \mu(0.3)$ and $y > x > normalized\ minutil = 0.3$

- For Cluster 2: $x = 0.27$ (lowest utility of cluster 2), $y = 0.27$ (highest utility of cluster 2)
  $|y - x| = |0.27 - 0.27| = 0 < \beta(0.1)$, $P\{x \le X \le y\} = 1/5 = 0.2 < \mu(0.3)$ and $y = x < normalized\ minutil = 0.3$
- For Cluster 3: $x = 0.6$ (lowest utility of cluster 3), $y = 0.6$ (highest utility of cluster 3)
  $|y - x| = |0.6 - 0.6| = 0 < \beta(0.1)$, $P\{x \le X \le y\} = 1/5 = 0.2 < \mu(0.3)$ and $y = x > normalized\ minutil = 0.3$

Since, the cluster 1 satisfies all the preliminaries, we can say that pattern *PQ* is a valid pattern and [0.33, 0.42] is the optimized interval of the utility. It is clearly seen that the interval of pattern *PQ* is optimized to [0.33, 0.42] from the original interval [0.27, 0.6]. Similarly, any collected pattern can be aggregated by clustering to check whether it contributes as an interesting pattern or not.

We design the algorithm for our aggregation model as: Let *PQ* be a collected pattern with utilities $util_1, util_2, util_3, \ldots, util_n$, from '*n*' different databases. '*minutil*' is the threshold set by the user, and μ and β are the thresholds set by experts.

Algorithm 1: AggregatingUtilities

```
1. for the utilities of pattern PQ do
     form dense clusters with adjacent utilities
2. for each cluster C do
     x = lowest utility in cluster C;
     y = highest utility in cluster C;
     diff = |y - x|;
     prob = |C|/n;
   endfor;
3. for each cluster C do
     if (diff ≤ β and prob ≥ μ and x ≥ normalized utility)
       output PQ is the valid pattern with utility interval [x, y];
     endif;
   endfor;
4. If there exists no cluster satisfying the preliminaries, then
     nutil = (nutil1+nutil2+ nutil3+...+nutiln)/n;
     if (nutil ≥ normalized minutil) then
       output PQ is the valid pattern with utility = nutil;
     endif;
   endif;
5. end.
```

The proposed algorithm aggregates the collected high-utility patterns from unknown and multiple databases by first putting the utilities in subspace for applying the normal distribution. The algorithm first checks if the clusters formed satisfy given thresholds to determine the interval of utility for every pattern. If the thresholds are not satisfied, then they are roughly aggregated by taking the average value of utilities.

5 Experimental Results

In this section, we evaluate the effectiveness and optimization of the utility interval based on the problem discussed in the previous section. We have implemented the proposed algorithm on the Intel core i3 processor with 8 GB of RAM. We have performed extensive experiments on three transaction databases taken from the SPMF repository [13].

5.1 Study 1

Retail is a sparse dataset containing 88,162 sequences of anonymous retail market basket data from an anonymous Belgian retail store having 16,470 distinct items. In experiments, we have used four partitions of databases from the retail dataset. Each partition with 22,040 transactions, is considered as a data source. We first mined all the partitions, each representing a data source, using the EFIM mining algorithm [14] and collected 5 common high-utility patterns with different utilities. We then applied our clustering-based model on these patterns to find the valid patterns and determine the optimized interval of their utility. The collected patterns from four unknown databases are shown in Table 6 with their respective utilities. The minimum utility threshold is set to 100,000 and the maximum utility in this transaction database is 481,021. Hence the *normalized minutil* is calculated to be 0.208 (minimum utility threshold/*Maximum utility*).

Table 6. Patterns collected from four databases

Pattern/Itemset	$util_1$	$util_2$	$util_3$	$util_4$	$util_5$
P_1 = {49,40}	113,324	125,182	118,774	123,741	95,675
P_2 = {49}	109,580	121,194	114,324	118,176	93,325
P_3 = {42,49}	102,765	74,223	73,905	-	-
P_4 = {42}	101,085	73,056	72,678	-	-
P_5 = {42,49}	97,234	76,344	74,716	-	-

After the patterns are aggregated by forming clusters, we got the following results:

- For pattern P_1, the clusters formed are cluster 1 = $nutil_1$, $nutil_2$, $nutil_3$ and $nutil_4$; cluster 2 = $nutil_5$. Since cluster 1 with $x = 0.236$, $y = 0.26$ and $y > x >$ *normalized minutil*, satisfies all the preliminaries so we can say that the pattern P_1 is a valid pattern with optimized utility interval [0.236, 0.26].

- For pattern P_2, the clusters formed are cluster 1 = $nutil_1$, $nutil_2$, $nutil_3$ and $nutil_4$; cluster 2 = $nutil_5$. Since cluster 1 with $x = 0.228$, $y = 0.252$ and $y > x >$ *normalized minutil*, satisfies all the preliminaries so we can say that the pattern P_2 is a valid pattern with optimized utility interval [0.228, 0.252].
- For pattern P_3, the cluster formed is cluster 1 = $nutil_1$, $nutil_2$ and $nutil_3$ with $x = 0.154$ and $y = 0.214$ and $y >$ *normalized minutil* $> x$. Hence, pattern P_3 is not a valid pattern.
- For pattern P_4, the cluster formed is cluster 1 = $nutil_1$, $nutil_2$ and $nutil_3$ with $x = 0.151$ and $y = 0.21$ and $y >$ *normalized minutil* $> x$. Hence, pattern P_4 is not a valid pattern.
- For pattern P_5, any of the utilities collected doesn't satisfy the minimum utility threshold (*minutil*) so we wipe out this pattern from global pattern analysis.

For both the valid patterns P_1 and P_2, the utility interval before aggregation and the optimized utility interval after aggregation are shown in Fig. 2.

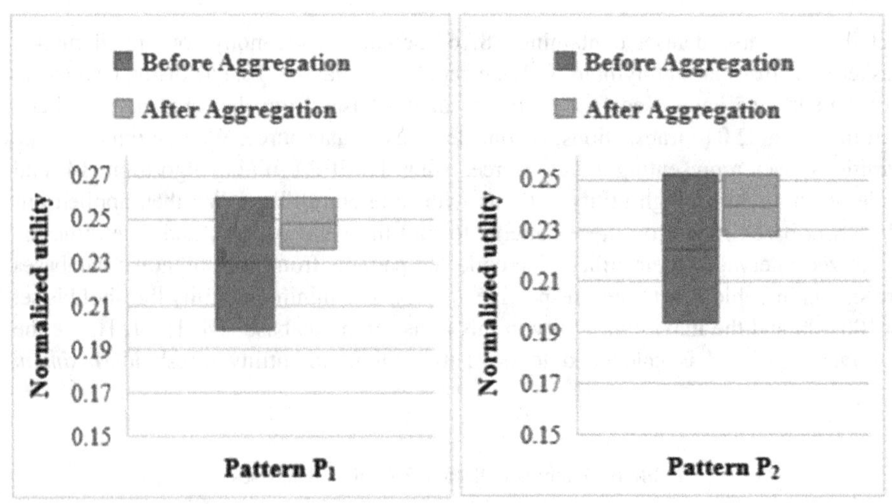

Fig. 2. Optimized utility interval for pattern P_1 and pattern P_2 (Retail)

5.2 Study 2

Kosarak is a very sparse dataset containing 990,000 sequences of click-stream data from a Hungarian news portal having 41,270 distinct items. In experiments, we have used five partitions of databases from the Kosarak dataset. Each partition with 198,000 transactions, is considered as a data source. We first mined all the partitions, each representing a data source, using the EFIM mining algorithm [14] and collected 3 common high-utility patterns with different utilities. We then applied our clustering-based model on these patterns to find the valid patterns and determine the optimized interval of their utility. The collected patterns from four unknown databases are shown in Table 7 with their respective utilities. The minimum utility threshold is set to 1,000,000 and the maximum utility in this transaction database is 17,824,032. Hence the *normalized minutil* is calculated to be 0.056 (minimum utility threshold/*Maximum utility*).

Table 7. Patterns collected from five databases

Pattern/Itemset	$util_1$	$util_2$	$util_3$	$util_4$	$util_5$
$P_1 = \{11,6\}$	3,544,830	3,594,574	3,591,954	3,599,700	3,492,974
$P_2 = \{148,218,11,6\}$	1,041,909	1,054,148	1,064,085	1,030,512	1030,090
$P_3 = \{218,11,6\}$	814,384	815,722	804,324	-	-

After the patterns are aggregated by forming clusters, we got the following results:

- For pattern P_1, the clusters formed are cluster 1 = $nutil_1$, $nutil_2$, $nutil_3$ and $nutil_4$; cluster 2 = $nutil_5$. Since cluster 1 with $x = 0.199$, $y = 0.202$ and $y > x >$ *normalized minutil*, satisfies all the preliminaries so we can say that the pattern P_1 is a valid pattern with optimized utility interval [0.199, 0.202].
- For pattern P_2, the clusters formed are cluster 1 = $nutil_1$, $nutil_2$, $nutil_4$ and $nutil_5$; cluster 2 = $nutil_3$. Since cluster 1 with $x = 0.058$, $y = 0.059$ and $y > x >$ *normalized minutil*, satisfies all the preliminaries so we can say that the pattern P_2 is a valid pattern with optimized utility interval [0.058, 0.059].
- For pattern P_3, any of the utilities collected doesn't satisfy the minimum utility threshold (*minutil*) so we wipe out this pattern from global pattern analysis.

For both the valid patterns P_1 and P_2, the utility interval before aggregation and the optimized utility interval after aggregation are shown in Fig. 3.

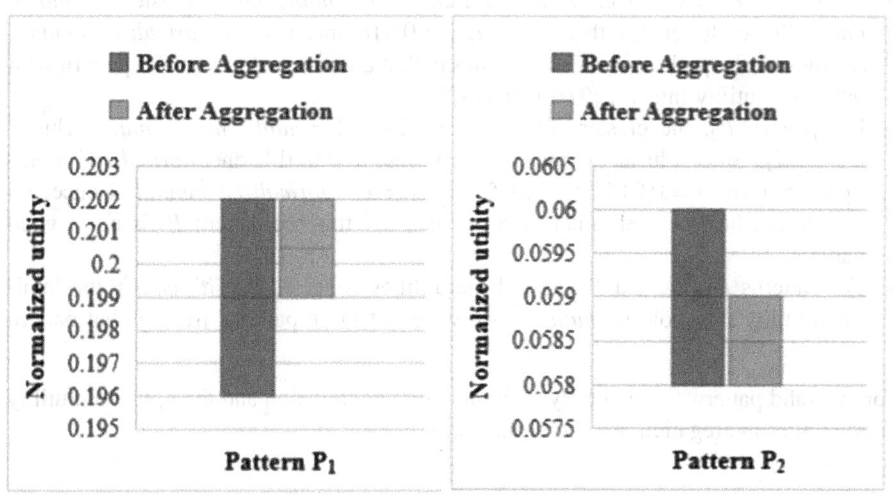

Fig. 3. Optimized utility interval for pattern P_1 and pattern P_2 (Kosarak)

5.3 Study 3

Chainstore is a very sparse dataset containing 1,112,949 sequences of customer transactions from a retail store, obtained and transformed from NU-Mine Bench having 46,086 distinct items. In experiments, we have used five partitions of databases from the Kosarak dataset. Each partition with 222,589 transactions, is considered as a data source. We first mined all the partitions, each representing a data source, using the EFIM mining algorithm [14] and collected 5 common high-utility patterns with different utilities. We then applied our clustering-based model on these patterns to find the valid patterns and determine the optimized interval of their utility. The collected patterns from four unknown databases are shown in Table 8 with their respective utilities. The minimum utility threshold is set to 1,200,000 and the maximum utility in this transaction database is 82,362,000. Hence the *normalized minutil* is calculated to be 0.015 (minimum utility threshold/*Maximum utility*).

Table 8. Patterns collected from five databases

Pattern/Itemset	$util_1$	$util_2$	$util_3$	$util_4$	$util_5$
P_1 = {39171,39688}	1,427,410	1,153,726	943,593	1,409,839	1,490,532
P_2 = {39143,39182}	1,156,200	726,000	748,200	1,243,200	-
P_3 = {21283,21308}	1,105,230	755,706	909,720	735,756	642,390
P_4 = {21297,21308}	610,869	500,745	758,100	-	-
P_5 = {39692,39690,39681}	538,692	508,521	-	-	-

After the patterns are aggregated by forming clusters, we got the following results:

- For pattern P_1, the clusters formed are cluster 1 = $nutil_1$, $nutil_2$; cluster 2 = $nutil_4$, $nutil_5$. Since cluster 2 with $x = 0.017$, $y = 0.018$, and $y > x >$ *normalized minutil*, satisfies all the preliminaries so we can say that the pattern P_1 is a valid pattern with optimized utility interval [0.017, 0.018].
- For pattern P_2, the clusters formed are cluster 1 = $nutil_1$, $nutil_2$, $nutil_3$; cluster 2 = $nutil_4$. Since cluster 1 has (a) $x = 0.009$, $y = 0.014$, and *normalized minutil* $> y > x$ (b) $x = 0.015$, $y = 0.015$, and $y = x =$ *normalized minutil*. Hence, no cluster satisfies the preliminaries so we can say that the pattern P_2 is not a valid pattern.
- For patterns P_3, P_4, and P_5, any of the utilities collected doesn't satisfy the minimum utility threshold (*minutil*) so we wipe out these patterns from global pattern analysis.

For the valid pattern P_1, the utility interval before aggregation and the optimized utility interval after aggregation are shown in Fig. 4.

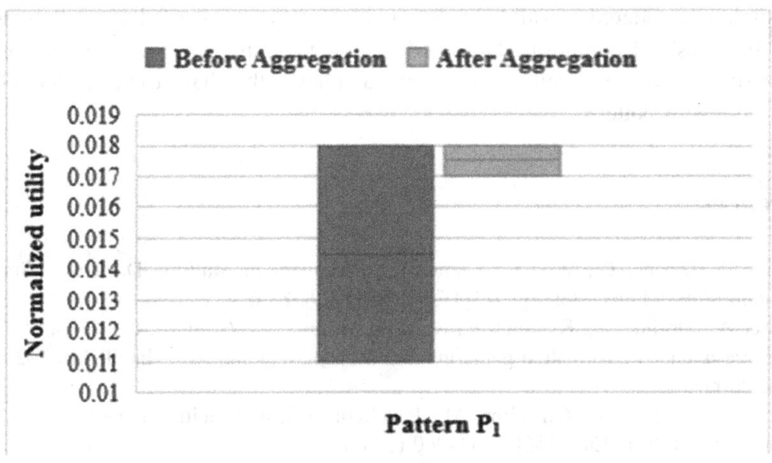

Fig. 4. Optimized utility interval for pattern P_1 (Chainstore)

5.4 Analysis

In the case of the retail dataset, the utility intervals for patterns P_1 and P_2 are optimized by 60% and 58% respectively. In the case of Kosarak dataset, the utility intervals for both patterns P_1 and P_2 are optimized by 50% whereas the utility for pattern P_1 in the Chainstore dataset is optimized by 85%. The databases considered in our experiments are large enough to establish that the cluster analysis performed on multiple databases gives promising results. Hence, the proposed method for parallel mining of unknown sources is able to aggregate the high-utility patterns in the union of a database of all the sources. The threshold β, which is nothing but an optimization degree plays a key role in deciding the amount of optimization to be done. In other words, it indicates the length of the interval to be optimized. The lower the value of threshold β, the higher will be the optimization degree. Higher optimization degree leads to more correct values of utility measure. Hence, in our experiments, we have tried to keep the value of β as low as possible to achieve higher optimization. The optimization degree also depends upon the nature of the database.

6 Conclusion

In this paper, we have effectively used the grid-based clustering method to form the clusters of collected patterns from unknown multiple databases. We used the CLIQUE technique to normally distribute the utilities in a subspace of all databases based on its adjacency values to other utilities. The method proposed identifies the valid patterns from the collection of all the mined patterns and also optimizes the utility interval based on the utility threshold. For convenience, we can take the lowest value of utility in the interval found. The experiments conducted on the dataset shows how the valid patterns are identified from a collection of patterns mined when the weighting model can't be applied to databases since its anonymity. The patterns which satisfy the threshold condition are considered as valid patterns and the rest are discarded when they do not satisfy the threshold condition. This model helps in finding out patterns mined from multiple sources whose identity is not known and hence they have to be clustered based on the utilities of patterns.

References

1. Zhang, S., Zhang, C., Wu, X.: Knowledge Discovery in Multiple Databases. Springer, London (2004). https://doi.org/10.1007/978-0-85729-388-6
2. Lesser, V., Horling, B., Klassner, F., Raja, A., Wagner, T., Zhang, S.X.: BIG: an agent for resource-bounded information gathering and decision making. Artif. Intell. **118**(1–2), 197–244 (2000)
3. Zhong, N., Yao, Y.Y., Ohishima, M.: Peculiarity oriented multi database mining. IEEE Trans. Knowl. Data Eng. **15**(4), 952–960 (2003)
4. Wu, X., Zhang, S.: Synthesizing high-frequency rules from different data sources. IEEE Trans. Knowl. Data Eng. **15**(2), 353–367 (2003)
5. Zhang, S., Zaki, M.J.: Mining multiple data sources: local pattern analysis. Data Min. Knowl. Discov. **12**(2–3), 121–125 (2006)
6. Adhikari, A., Ramachandra Rao, P., Pedrycz, W.: Developing Multi-Database Mining Applications. Springer, London (2010). https://doi.org/10.1007/978-1-84996-044-1
7. Muley, A., Gudadhe, M.: Synthesizing high-utility patterns from different data sources. Data. **3**(3), 32 (2018)
8. Arabie, P., Hubert, L.J.: An overview of combinatorial data. In: Clustering and Classification, p. 5 (1996)
9. Piatetsky-Shapiro, G., Fayyad, U.M., Smyth, P., Uthurusamy, R. (eds.) Advances in Knowledge Discovery and Data Mining, vol. 21. AAAI Press, Menlo Park (1996)
10. Michalski, R.S., Stepp, R.E.: Learning from observation: conceptual clustering. In: Michalski, R.S., Carbonell, J.G., Mitchell, T.M. (eds.) Machine Learning. SYMBOLIC, vol. 1, pp. 331–363. Springer, Berlin (1983). https://doi.org/10.1007/978-3-662-12405-5_11
11. Han, J., Pei, J., Kamber, M.: Data Mining: Concepts and Techniques. Elsevier, New York (2011)
12. Agrawal, R., Gehrke, J., Gunopulos, D., Raghavan, P.: Automatic subspace clustering of high dimensional data for data mining applications. ACM SIGMOD Rec. **27**(2), 94–105 (1998)
13. Fournier-Viger, P., et al.: The SPMF open-source data mining library version 2. In: Berendt, B., et al. (eds.) ECML PKDD 2016. LNCS (LNAI), vol. 9853, pp. 36–40. Springer, Cham (2016). https://doi.org/10.1007/978-3-319-46131-1_8
14. Zida, S., Fournier-Viger, P., Lin, J.C.W., Wu, C.W., Tseng, V.S.: EFIM: a fast and memory efficient algorithm for high-utility itemset mining. Knowl. Inf. Syst. **51**(2), 595–625 (2017)
15. Good, I.: Probability and the Weighting of Evidence. Charles Griffin, London (1950)
16. Chen, Y., An, A.: Approximate parallel high-utility itemset mining. Big Data Res. **6**, 26–42 (2016)
17. Xun, Y., Zhang, J., Qin, X.: FiDoop: parallel mining of frequent itemsets using mapreduce. IEEE Trans. Syst. Man Cybern.: Syst. **46**(3), 313–325 (2016)
18. Zhang, F., Liu, M., Gui, F., Shen, W., Shami, A., Ma, Y.: A distributed frequent itemset mining algorithm using spark for big data analytics. Clust. Comput. **18**(4), 1493–1501 (2015)

19. Marjani, M., et al.: Big IoT data analytics: architecture, opportunities, and open research challenges. IEEE Access **5**, 5247–5261 (2017)
20. Wang, R., et al.: Review on mining data from multiple data sources. Pattern Recognit. Lett. **109**, 120–128 (2018)
21. Adhikari, A., Adhikari, J.: Mining patterns of select items in different data sources. Advances in Knowledge Discovery in Databases. ISRL, vol. 79, pp. 233–254. Springer, Cham (2015). https://doi.org/10.1007/978-3-319-13212-9_12
22. Agrawal, R., Imieliński, T., Swami, A.: Mining association rules between sets of items in large databases. ACM SIGMOD Rec. **22**(2), 207–216 (1993)
23. Yao, H., Hamilton, H.J., Geng, L.: A unified framework for utility-based measures for mining itemsets. In: Proceedings of ACM SIGKDD 2nd Workshop on Utility-Based Data Mining, pp. 28–37, August 2006
24. Ahmed, C.F., Tanbeer, S.K., Jeong, B.S., Lee, Y.K.: Efficient tree structures for high-utility pattern mining in incremental databases. IEEE Trans. Knowl. Data Eng. **21**(12), 1708–1721 (2009)
25. Fournier-Viger, P., Wu, C.-W., Tseng, Vincent S.: Novel concise representations of high utility itemsets using generator patterns. In: Luo, X., Yu, J.X., Li, Z. (eds.) ADMA 2014. LNCS (LNAI), vol. 8933, pp. 30–43. Springer, Cham (2014). https://doi.org/10.1007/978-3-319-14717-8_3
26. Lin, Y., Chen, H., Lin, G., Chen, J., Ma, Z., Li, J.: Synthesizing decision rules from multiple information sources: a neighborhood granulation viewpoint. Int. J. Mach. Learn. Cybern. **9**, 1919–1928 (2018)
27. Zhang, S., Wu, X., Zhang, C.: Multi-database mining. IEEE Comput. Intell. Bull. **2**(1), 5–13 (2003)
28. Xu, W., Yu, J.: A novel approach to information fusion in multi-source datasets: a granular computing viewpoint. Inf. Sci. **378**, 410–423 (2017)

A Study of Three Different Approaches to Point Placement on a Line in an Inexact Model

Kishore Kumar V. Kannan, Pijus K. Sarker, Amangeldy Turdaliev, Asish Mukhopadhyay(✉), and Md. Zamilur Rahman

School of Computer Science, University of Windsor, Windsor, ON N9B 3P4, Canada
{varadhak,sarkerp,turdalia,asishm,rahma11u}@uwindsor.ca

Abstract. The point placement problem is to determine the locations of n distinct points on a line uniquely (up to translation and reflection) by making the fewest possible pairwise distance queries of an adversary (an adversary is just a source of true distances). A number of deterministic and randomized algorithms are available when distances are known exactly. In this paper, we discuss the problem in an inexact model. This is when distances returned by the adversary are not exact; instead, only upper and lower bounds on the distances are provided. We explore three different approaches to this problem. The first is based on an adaption of a distance geometry approach that Havel and Crippen [6] used to solve the molecular conformation problem. The second is based on a linear programming solution to a set of difference constraints that was used by Mumey [7] to solve a probe location problem arsing in DNA sequence analysis. The third is based on a heuristic called Stochastic Proximity Embedding, proposed by Agrafiotis [8]. Extensive experiments were carried out to determine the most promising approach vis-a-vis two parameters: run-time and quality of the embedding, measured by computing a certain stress function.

Keywords: Probe location problem · Distance geometry · Point placement · Bound smoothing · Embed algorithm · Eigenvalue decomposition

1 Introduction

Background: The point-placement problem can be described thus: Determine the locations of n distinct points on a line from pairwise distance queries. The goal is to minimize the number of queries. The queries are batched, with the queries in the i-th batch depending on the queries of the previous batches. The queries in a batch are presented as a query graph, whose edges join pairs of points whose distances are being queried. Each batch query is a round and k query batches define a k-round algorithm.

Variants of this problem have been studied in a surprisingly diverse number of areas: computational biology, learning theory, computational geometry, etc.

M. L. Gavrilova and C. J. K. Tan (Eds.): Trans. on Comput. Sci. XXXIV, LNCS 11820, pp. 44–63, 2019.
https://doi.org/10.1007/978-3-662-59958-7_3

In learning theory [3] this problem is posed as one of learning a set of points on a line non-adaptively, when learning has to proceed based on a fixed set of given distances, or adaptively when learning proceeds in rounds, with the edges queried in one round depending on those queried in the previous rounds.

In computational geometry the problem is studied as the turnpike problem. The description is as follows. On an expressway stretching from town A to town B there are several gas exits; the distances between all pairs of exits are known. The problem is to determine the geometric locations of these exits. This problem was first studied by Skiena et al. [9] who proposed a practical heuristic for the reconstruction.

In computational biology, it appears in the guise of the restriction site mapping problem. Biologists discovered that certain restriction enzymes cleave a DNA sequence at specific sites known as restriction sites. For example, it was discovered by Smith and Wilcox [10] that the restriction enzyme Hind II cleaves DNA sequences at the restriction sites GTGCAC or GTTAAC. In lab experiments, by means of fluorescent in situ hybridization (FISH experiments) biologists are able to measure the lengths of such cleaved DNA strings. Given the distances (measured by the number of intervening nucleotides) between all pairs of restriction sites, the task is to determine the exact locations of the restriction sites.

The turnpike problem and the restriction mapping problem are identical, except for the unit of distance involved; in both of these we seek to fit a set of points to a given set of inter-point distances. As is well-known, the solution may not be unique and the running time is polynomial in the number of points.

While the point placement problem, prima facie, bears a resemblance to these two problems it is different in its formulation - we are allowed to make pairwise distance queries among a distinct set of labeled points. It turns out that it is possible to determine a unique placement of the points up to translation and reflection in time that is linear in the number of points.

Many interesting results on 1-round and 2-round algorithms were discovered by Damaschke [3]; these results were refined and extended by Chin et al. [2]. State of the art results for 2-round algorithms are to be found in [11]. At the heart of all such algorithms lies the problem of designing a suitable query graph which is constructed by gluing together several copies of a small graph which is either line-rigid or can be made so by adding suitable conditions on the edge lengths of this graph that make it line-rigid. The set of rigidity conditions can be largely, and finding thess requires intricate analysis. The successive rounds are exploited to meet these rigidity conditions.

The concept of line-rigidity deserves a formal definition. A graph is said to be line-rigid if the lengths assigned to its edges define a unique placement of the points, up to translation and reflection. Such an assignment of lengths is said to be valid. A decision algorithm to test for validity is discussed later in this introduction.

The simplest point-placement algorithm that takes only 1-round is the 3-cycle algorithm. The nomenclature is due to the fact that we glue together several copies of a 3-cycle (triangle) as shown in Fig. 1 below. The placement is solved in one round as a triangle is line-rigid for a valid assignment of lengths for its edges, which is when the sum or difference of the lengths of two of the edges is equal to the length of the third.

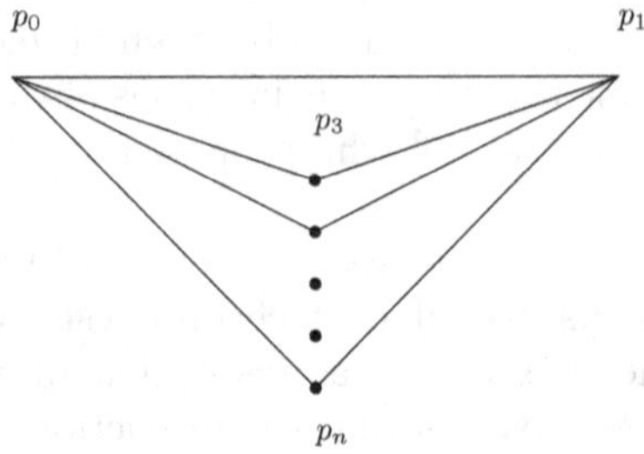

Fig. 1. Query graph using triangles

The point-placement problem is a special case of a well-known distance geometry problem : recover the coordinates of a set of points $P = \{p_0, p_1, p_2, \ldots, p_{n-1}\}$, given a set of mutual distances. When all pairwise distances are known, Young and Householder [12] showed that a necessary and sufficient condition is that the matrix $B = [b_{ij}] = [(d_{i0}^2 + d_{j0}^2 - d_{ij}^2)/2]$ is positive semi-definite, where d_{ij} is the Euclidean distance between p_i and p_j. The rank r of B also gives the minimum dimension of the Euclidean space in which the point set P can be embedded. A matrix which involves the square of the distances more symmetrically is the symmetric $(n+1) \times (n+1)$ Cayley-Menger matrix, defined by $C_{0i} = C_{i0} = 1$ for $0 < i \leq n$, $C_{00} = 0$ and $C_{ij} = D_{ij}$ for $1 \leq i, j \leq n$, [12]. It can be shown that the rank of the Cayley-Menger matrix C is $r + 2$. We can now formally define the notion of validity of an assignment of lengths for a point placement problem. This is when the rank of C is at most 3 (this is a special case of the result that there exists a r-dimensional embedding of the query graph if the rank of C is at most r+2; our claim follows by setting $r = 1$, see [13]).

A prototypical point-placement problem was introduced by Saxe [14]. Given a weighted graph $G = (V, E, w)$, where V and E are the set of its vertices and edges respectively and w is a non-negative real-valued function on the set of edges, does there exist an embedding of the vertices V in an Euclidean space E^k such that the weight of an edge is the Euclidean distance between the pair of points in the embedded space, corresponding to the vertices that the edge connects. This problem is strongly NP-complete even when $k = 1$ and the distances are restricted to the set $\{1, 2, 3, 4\}$. It is, however, possible to solve this problem for special classes of graphs when the embedding dimension is 1 [15]. The edges of such graphs join pairs of points for which the distances are known.

In this paper, we study a one-round version of the the point placement problem in the adversarial model when the query distances returned by the adversary are not necessarily exact. A formal definition of the problem goes as follows.

Problem Statement: Let $P = \{p_0, p_1, ..., p_{n-1}\}$ be a set of n distinct points on a line. For a pair of points p_i and p_j, the query distance $D(p_i, p_j)$ lies in an interval $[l_{ij}, u_{ij}]$, where $l_{ij} \leq u_{ij}$ denote the lower and upper bound respectively. The objective is to find a set of locations of the points in P that are consistent with the given distance bounds.

The distance bounds are collectively represented by an upper distance matrix, U, for the upper bounds and a lower distance matrix, L, for the lower bounds. The corresponding entries in the upper and lower bound matrices for pairs of points for which $l_{ij} = u_{ij}$, that is, the pairwise distances are exactly known, are identical. When the distance between a pair of points is unknown, the distance interval bound is set to $[-\infty, \infty]$.

For example with three points ($n = 3$), a possible pair of upper and lower bound distance matrices are shown below:

$$U(p_0, p_1, p_2) = \begin{pmatrix} 0 & 60 & \infty \\ 60 & 0 & 3 \\ \infty & 3 & 0 \end{pmatrix}$$

$$L(p_0, p_1, p_2) = \begin{pmatrix} 0 & 60 & -\infty \\ 60 & 0 & 3 \\ -\infty & 3 & 0 \end{pmatrix}$$

Thus, $l_{01} = u_{01} = 60$ for the pair of points p_0 and p_1, whereas $l_{02} = -\infty$ and $u_{02} = \infty$ for the pair p_0 and p_2.

Motivation: In [5,16] we implemented a number of 1-round, 2-round and 3-round algorithms and compared their performances with respect to query complexity (upper bound on the number of queries made by the algorithms) and time complexity (running times of the algorithms). Distance queries were returned based on an apriori knowledge of a linear layout of n points with integral coordinates and the correctness of the outputs of our programs were tested by checking the apriori layouts with the computed layouts. Exact arithmetic facilitated the checking of the rigidity conditions. Our experience with the implementations drew our attention to the difficulty of checking the rigidity conditions when the coordinates of the points are arbitrary real numbers and to the importance of solving the problem in an inexact model.

Further motivation comes from the fact that in many applications distances are not known exactly, only that they lie between some bounds and thus fits into the above model. Such is the case for the probe location problem in computational biology, where distances between some pairs of probes are not known exactly. This problem is the focus of this paper. A similar situation arises in the molecular conformation problem from computational chemistry [17]. In this case distances between some pairs of atoms are known to lie between some bounds and we are required to find all possible conformations of the molecules consistent with the distance bounds. A third application is the computation of genetic linkage maps that has been studied in [18].

Overview of Contents: The rest of the paper is organized thus. In the next section, we show how the point placement problem in the inexact model can be formulated in the distance geometry framework. Based on this formulation, we propose an algorithm, called DGPL (short for Distance Geometry based Probe Location). In the third section, we briefly review an algorithm by Mumey [7] for the probe location problem, an abstract formulation of which is the point location problem in the inexact model. In the fourth section, we discuss a novel heuristic called Stochastic Proximity Embedding that applies to the one-dimensional point placement problem. In the next section, experimental results are given, comparing the performance of the DGPL algorithm with that of Mumey's and a thorough experimental study of the SPE heuristic. In the final section, we conclude with some observations and directions for further research and discuss the applicability of the DGPL technique to determining the conformations of a molecule.

2 The Distance Geometry Approach

The fundamental problem of distance geometry is this: "Given a set of m distance constraints in the form of lower and upper bounds on a (sparse) set of pairwise distance measures, and chirality constraints on quadruples of points, find all possible embeddings of the points in a suitable k-dimensional Euclidean space." [6] (see also [1,4]).

Crippen and Havel's EMBED algorithm for the molecular conformation problem is based on a solution to the above fundamental problem. Let d_{ij} denote the distance between a pair of points p_i and p_j; then each of the m constraints above specifies an upper and lower bound on some d_{ij}. For our 1-dimensional point placement problem, chirality constraints do not come into the picture. Determining a feasible set of locations of the set of points in P is equivalent to finding the coordinates of these locations, with reference to some origin of coordinates. Thus the approach used for the EMBED algorithm can be directly adapted to the solution of our problem. Due to the non-availability of an implementation of the EMBED algorithm, we wrote our own.

Below, we describe the DGPL algorithm, on which our implementation is based. Before this, the so-called bound-smoothing algorithm used in the EMBED algorithm deserves some mention. The main underlying idea is to refine distance bounds into distance limits. Thus if $[l'_{ij}, u'_{ij}]$ are the distance limits corresponding to distance bounds $[l_{ij}, u_{ij}]$, then $[l'_{ij}, u'_{ij}] \subset [l_{ij}, u_{ij}]$. The distance limits are defined as follows.

Let l_{ij} and u_{ij} be the upper and lower distance bounds on d_{ij}. Then

$$l'_{ij} = \inf_e \{d_{ij} | l_{ij} \leq d_{ij} \leq u_{ij}\}$$

and

$$u'_{ij} = \sup_e \{d_{ij} | l_{ij} \leq d_{ij} \leq u_{ij}\},$$

where the inf and sup are taken over all possible embeddings $e : P \times P \rightarrow R$ of the points in P on a line.

When this is done, assuming that the triangle inequality holds for the triplet of distances among any three points p_i, p_j, p_k, then these distance limits are called triangle limits.

In [19], a characterization of the triangle limits was established. It was also shown how this could be exploited to compute the triangle limits, using a modified version of Floyd's shortest path algorithm [20].

2.1 Algorithm DGPL

An input to the DGPL program for a set of n points in its most general form consists of the following: (a) exact distances between some pairs of points; (b) finite upper and lower bounds for some other pairs; (c) for the rest of the pairs, the distance bounds are not specified. However, as we have conceived of DGPL as a solution to the point placement problem in the inexact model, the input is assumed to be adversarial and is simulated by generating an initial adversarial layout of the points. This layout is used by the algorithm to generate input data of types (a) and (c); however, DGPL will also work if some of the input data are of type (b).

The DGPL algorithm works in three main phases, as below.

Phase 1: [Synthetic Data Generation] Based on the user-input for n, the algorithm simulates an adversary to create a layout of n points; further, based on the user-input for the number of pairs for which mutual distances are not known, the algorithm assigns the distance bounds $[-\infty, \infty]$ for as many pairs and uses the layout created to assign suitable distance bounds for the rest of the pairs; the output from this stage are the lower and upper bound distance matrices, L and U.

Phase 2: [Point location or Coordinate generation] A set of coordinates of the n points, consistent with the bounds in the matrices L and U, are computed, following the approach in [17] for the molecular conformation problem.

Phase 3: [Layout generation] Finally, a layout of the embedding is generated, which allows us to verify how good the computed layout is vis-a-vis the adversarial layout.

A more detailed description of the algorithm is given below.

In the next section, we discuss the simulation of DGPL for a small value of n.

Algorithm 1. DGPL

Input: 1. The size n of P.
2. The number of pairs for which $l = -\infty$ and $u = \infty$.
3. Embedding dimension

Output: Locations of the n points in P, consistent with the distance bounds

(1) Create a random valid layout of the points p_i in $P = \{p_0, p_1, ..., p_{n-1}\}$.

(2) Create distance bound matrices L and U.

(2.1) Assign $-\infty$ and ∞ respectively to the corresponding values in the L and U matrices for as many pairs as the the number unknown distances (user-specified).

(2.2) Assign the distances determined from the layout of Step 1 to the remaining entries of both L and U

(3) Apply a modified version of Floyd's shortest path algorithm [18] to compute triangle limit matrices, LL and UL, from the distance bound matrices, L and U.

(4) If all triangle limit intervals have been collapsed to a single value (which is now the distance between the corresponding pair of points) go to Step 5, else collapse any remaining pair of triangle limit interval to a randomly chosen value in this interval and go to Step 3 (this step is called metrization)

(5) Compute matrix $B = [b_{ij}] = [(d_{i0}^2 + d_{j0}^2 - d_{ij}^2)/2]$, where d_{ij} is the distance between points p_i and p_j, $1 \leq i, j \leq n-1$ [13].

(6) Compute the eigenvalue decomposition of the matrix B; the product of the largest eigenvalue with its corresponding normalized eigenvector gives the one-dimensional coordinates of all the points.

(7) The computed coordinates are plotted on a line, placing p_0 at the origin of coordinates, and are compared with the adversarial layout for the extent of agreement.

2.2 Simulating DGPL

1. Let the input to the program be as follows:
   Number of points: 4
   Number of unknown distances: 1
   Embedding dimension: 1.
2. The following random one-dimensional layout of the four points is created:

$$p_0 = 0, p_1 = 59, p_2 = 48, p_3 = 74.$$

3. The input lower and upper bound distance matrices, L and U, are set to:

$$L = \begin{pmatrix} 0 & 59 & 48 & 74 \\ 59 & 0 & 11 & -\infty \\ 48 & 11 & 0 & 26 \\ 74 & -\infty & 26 & 0 \end{pmatrix}, U = \begin{pmatrix} 0 & 59 & 48 & 74 \\ 59 & 0 & 11 & \infty \\ 48 & 11 & 0 & 26 \\ 74 & \infty & 26 & 0 \end{pmatrix},$$

   where the distance d_{13} between the points p_1 and p_3 is contained in the interval $[-\infty, \infty]$.

4. The lower and upper triangle limit matrices, LL and UL, obtained by bound smoothing from the lower and upper distance bound matrices are:

$$LL = \begin{pmatrix} 0 & 59 & 48 & 74 \\ 59 & 0 & 11 & 15 \\ 48 & 11 & 0 & 26 \\ 74 & 15 & 26 & 0 \end{pmatrix}, UL = \begin{pmatrix} 0 & 59 & 48 & 74 \\ 59 & 0 & 11 & 37 \\ 48 & 11 & 0 & 26 \\ 74 & 37 & 26 & 0 \end{pmatrix}.$$

5. The distance limits are then metrized, as discussed in Step 4 of the DGPL algorithm, to fixed distances as:

$$D = \begin{pmatrix} 0 & 59 & 48 & 74 \\ 59 & 0 & 11 & 35 \\ 48 & 11 & 0 & 26 \\ 74 & 35 & 26 & 0 \end{pmatrix}.$$

6. Relative to the point p_0 as the origin of coordinates, the matrix B computes to:

$$B = \begin{pmatrix} 3481 & 2832 & 3866 \\ 2832 & 2304 & 3552 \\ 3866 & 3552 & 5476 \end{pmatrix}.$$

7. From the eigenvalue decomposition of the matrix B, we obtain the following coordinate matrix as the product of the square root of largest eigenvalue with its corresponding normalized eigenvector:

$$X = \begin{pmatrix} 58.677 \\ 48.828 \\ 73.671 \end{pmatrix}.$$

The 1-dimensional embedding of the computed coordinates, after rounding to integer values, is shown in Fig. 2; the agreement with the adversarial layout is seen to be near-perfect.

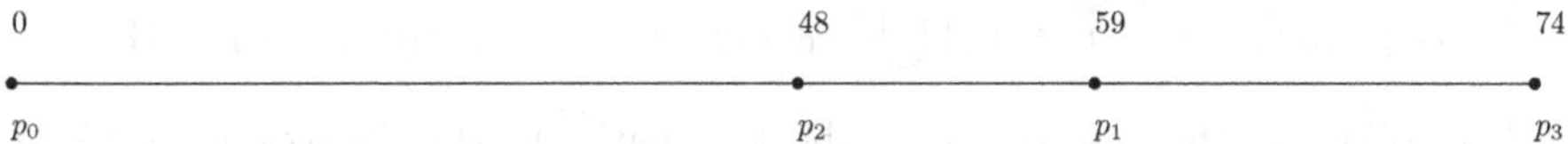

Fig. 2. Final embedding of the four input points

3 Mumey's Approach to the Probe Location Problem

Mumey [7] considered the problem of mapping probes along a chromosome based on separation or distance intervals between probe pairs, estimated from *fluorescence in-situ hybridization (FISH)* experiments. He named this as the probe location problem. The problem is challenging as the distance intervals are known only with some confidence level, some may be error-prone and these need to be identified to find a consistent map. Previous algorithmic approaches based on seriation [21], simulated annealing [22] and branch and bound algorithm [23], relying as these did on exhaustive search, could handle inputs of only up to 20 or fewer probes efficiently. However, Mumey's algorithm claims to be able solve the problem for up to 100 probes in a matter of few minutes. We briefly review Mumey's algorithm next.

From the above discussion it is clear that the probe location problem can also be conveniently cast in the framework of distance geometry as a concrete version of a point placement problem in the inexact model. Indeed, we have done this and the results are discussed in the next section.

3.1 Overview of Mumey's Algorithm

Let $P = \{p_0, p_1, ..., p_{n-1}\}$ be the list of probes in a chromosome, it being given that the distance interval between a pair of probes p_i and p_j lie in an interval $[l_{ij}, u_{ij}]$, where l_{ij} and u_{ij} are respectively the lower and upper bounds on the distance. The probe location problem is to identify a feasible set of probe locations $\{x_0, x_1, ..., x_{n-1}\}$ from the given distance intervals such that $|x_i - x_j| \in [l_{ij}, u_{ij}]$.

The distance bounds on probe pairs leads to a special kind of linear program, which is a system of difference constraints. A directed graph derived from these constraints is input to Bellman-Ford's single source shortest path algorithm. The shortest paths $\delta(v_0, v_i)$, $1 \leq i \leq n$, from the specially-added source vertex, v_0, to all the destination vertices, v_i, $i = 1, \ldots, n$ is a solution to the system of difference constraints [20].

Let m difference constraints over n variables be defined by

$$x_i - x_j \leq b_k,$$

where $1 \leq i, j \leq n$ and $1 \leq k \leq m$. We can abbreviate this as a linear program $Ax \leq b$. The *constraint graph*, $G = (V, E)$, corresponding to this linear program is defined by the vertex set $V = \{v_0, v_1, \ldots, v_n\}$, and edge set

$$E = \{(v_i, v_j) : x_j - x_i \leq b_k\} \bigcup \{(v_0, v_1), (v_0, v_2), (v_0, v_3), \ldots, (v_0, v_n)\}.$$

The following theorem connects shortest paths in the constraint graph G from the source vertex v_0, with a solution of the linear program $Ax \leq b$.

Theorem 1 [20]. *Given a system $Ax \leq b$ of difference constraints, let $G = (V, E)$ be the corresponding constraint graph. If G contains no negative-weight cycles, then*

$$x = (\delta(v_0, v_1), \delta(v_0, v_2), \ldots, \delta(v_0, v_n)\}$$

is a feasible solution for the system, If G contains a negative-weight cycle, then there is no feasible solution.

For the probe location problem, the directed graph in question is called an edge orientation graph, whose construction is described next.

Edge Orientation Graph. The first step in the construction of an edge orientation graph is to set the orientation of each edge by choosing one placement. If x_i and x_j are the positions of probes i and j then one of the following is true:

$$x_j - x_i \in [l_{ij}, u_{ij}] \tag{1}$$

$$x_j - x_i \in [l_{ij}, u_{ij}] \tag{2}$$

If (1) holds then x_i is to the left of x_j whereas if (2) holds, then x_j is to the left of x_i. Assume that (1) holds. We can express this equivalently as.

$$l_{ij} \leq x_j - x_i \leq u_{ij}$$

or as:

$$x_j - x_i \leq u_{ij} \tag{3}$$

$$x_i - x_j \leq -l_{ij} \tag{4}$$

Corresponding to the two inequalities above, we have two edges in the edge orientation graph, one going from x_i to x_j, with weight u_{ij} and the other from x_j to x_i with weight $-l_{ij}$.

Similarly, if (2) holds, we can express this equivalently as:

$$l_{ij} \leq x_i - x_j \leq u_{ij}$$

or as:

$$x_i - x_j \leq u_{ij} \tag{5}$$

$$x_j - x_i \leq -l_{ij} \tag{6}$$

with two directed edges in the edge orientation graph of weights u_{ij} and $-l_{ij}$ respectively.

When $l_{ij} = u_{ij}$ for a pair of probes p_i and p_j, there is exactly one edge connecting the corresponding nodes in the edge orientation graph, its orientation determined by the relative linear order of the probes.

Finding Feasible Probe Positions. Once all the edge weights are fixed, a special source vertex is added whose distances to all other vertices are initialized to 0 and Bellman-Ford's algorithm is run on this modified edge orientation graph. If there is no negative cycle in the graph, Bellman-Ford's algorithm outputs a feasible solution $(x_0, x_1, ..., x_{n-1})$. Note that $(x_0+d, x_1+d, ..., x_{n-1}+d)$, for any constant d is also a feasible solution. Indeed the solution to a point placement problem is unique only up to translation. Otherwise, the algorithm resets the edge weights by changing the relative order of a probe pair and re-runs the Bellman-Ford algorithm. These two steps are repeated till a feasible solution is found.

4 Stochastic Proximity Embedding

In contrast to the rather elaborate approaches of DGPL and Mumey, a novel and extremely simple heuristic, called Stochastic Proximity Embedding (SPE) proposed by Agrafiotis in [8], is applicable to the one-dimensional point placement problem. It is worth noting that it is applicable to higher-dimensional problems as well, including the molecular conformation problem we discussed in the last section. This heuristic was originally proposed as a faster alternative to standard techniques like Multi-Dimensional Scaling (MDS) and Principal Component Analysis (PCA) that are used to recover the coordinates of n points in a k-dimensional Euclidean space R^k from their mutual distances in $O(n^2)$ time.

The input to SPE is a partial or complete distance matrix, $R = [r_{ij}]$, that correspond to the unknown embedding of the actual points $P = \{p_1, p_2, \ldots, p_n\}$ that yield these distances. It starts with an arbitrary embedding $P^a = \{{p_1}^a, {p_2}^a, \ldots, {p_n}^a\}$ in R^k, and $D = [d_{ij}]$, the distance matrix obtained from this embedding.

The SPE heuristic has two cycles: an inner cycle nested inside an outer cycle. The latter is the learning cycle, while the inner cycle picks up pairs of points randomly from the set P^a. Each randomly picked pair is subjected to a Newton-Raphson root-finding style of correction to the current coordinates. The number of iterations of the algorithm is $C * S$, where C is the number of steps for the learning cycle, and S is the number of times a random pair of points is selected from the point set P^a. The parameters C and S are set so that $CS = o(n^2)$. The output of SPE are the coordinates of the point set at the termination of the learning cycle.

Algorithm 2. SPE [8]

Input: A random embedding of points in R^k
Output: Updated coordinates of the same points, consistent with the R-matrix

Initialize the coordinates of P^a
$k \leftarrow 1$
while $(k \leq C)$ **do**
    $l \leftarrow 1$
    while $(l \leq S)$ **do**
        For a random pair of points, p_i^a and p_j^a, compute distance $d_{ij} = ||p_i^a - p_j^a||$
        if $(d_{ij} \neq r_{ij})$ **then**
            $p_i^a \leftarrow p_i^a + \lambda \frac{1}{2} \frac{r_{ij} - d_{ij}}{d_{ij} + \epsilon} (p_i^a - p_j^a)$, $\epsilon \neq 0$
            $p_j^a \leftarrow p_j^a + \lambda \frac{1}{2} \frac{r_{ij} - d_{ij}}{d_{ij} + \epsilon} (p_j^a - p_i^a)$, $\epsilon \neq 0$
        end if
        $l \leftarrow l + 1$
    end while
    $\lambda \leftarrow \lambda - \delta\lambda$
    $k \leftarrow k + 1$
end while

5 Experimental Results

We implemented both the DGPL algorithm and Mumey's, discussed in the previous sections in Python 2.7. The programs were run on a computer with the following configuration: Intel(R) Xeon(R) CPU, X7460@2.66 GHz OS: Ubuntu 12.04.5, Architecture:i686. We used the mathematical package numpy.linAlg to calculate eigenvalue decompositions and also for solving linear equations; the package matplotlib.pyplot was used to plot an embedding of final coordinates obtained from the programs in a space of specified dimensionality.

The chart below compares the running time of Mumey's algorithm with that of DGPL. Each of these algorithms were run on point sets of different sizes, up to 101 points. We also recorded the effect on both algorithms of the number of pairs of points (probes) for which the distances are not known or are unspecified. In these cases, we set $l_{ij} = -\infty$ and $u_{ij} = \infty$ (Table 1).

Table 1. Performance comparison of Mumey's and the DGPL algorithm

No. of points	No. of unknown distances	Mumey's approach running time (hrs:mins:secs)	DGPL algorithm running time (hrs:mins:secs)
3	1	0:00:00.000184	0:00:00.001514
10	2	0:00:00.001339	0:00:00.006938
10	5	0:00:00.024560	0:00:00.006816
10	8	0:00:00.060520	0:00:00.017163
20	2	0:00:00.001369	0:00:00.007464
20	5	0:00:00.001336	0:00:00.007743
20	10	0:00:01.164363	0:00:00.007436
40	5	0:00:00.947250	0:00:00.328563
40	8	0:00:07.369925	0:00:00.315001
40	10	0:00:30.857658	0:00:00.312674
80	5	0:00:10.609233	0:00:02.503798
80	10	0:06:15.443501	0:00:02.496285
80	15	5:00:00.000000+	0:00:02.687672
101	5	0:00:14.256343	0:00:05.020695
101	10	0:10:32.299084	0:00:05.282747
101	15	5:00:00.000000+	0:00:05.192594

Interestingly, the chart (Fig. 3) shows that Mumey's approach takes a longer time when the number of unknown distances increases. Each of these algorithms were run on point sets of different sizes, up to 100 points.

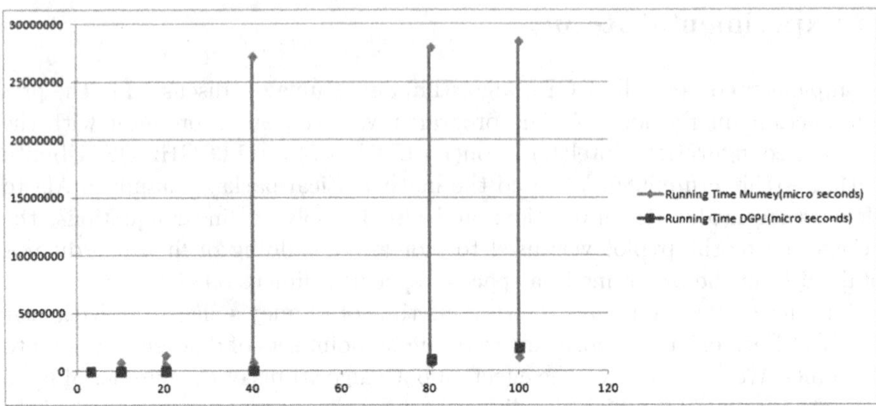

Fig. 3. Time complexity graph Mumey's vs DGPL algorithm - Increasing number of unknown distances between fixed number of points

Clearly, the DGPL algorithm is consistently faster; as we can see from the graph, irrespective of the number of unknown distances in the fixed number of points, DGPL's run-time is linear in the number of points. However, this is not true of Mumey's approach. This can be explained by the fact that when a negative cycle is detected Bellman-Ford's algorithm has to be run again. This is more likely to happen as the number of unknown distances increase. Thus the running time increases rapidly. Furthermore, after each detection of a negative cycle, a new distance matrix will have to be plugged into the Bellman-Ford algorithm to find the feasible solutions. In addition, the cost of keeping track of the distances chosen for unknown distance pairs increases with the number of such pairs.

The quality of the embedding produced by SPE is measured using the following stress function as described in [8]

$$S = \frac{\Sigma_{i<j} \frac{(d_{ij}-r_{ij})^2}{r_{ij}}}{\Sigma_{i<j} r_{ij}} \tag{7}$$

In the inexact model, distances between pairs of points are only known up to an interval defined by distance limits. Let I_{ij} be the distance interval, corresponding to a pair of points p_i and p_j. The r_{ij} in line 7 of the original SPE Algorithm 2 is chosen in two different ways.

1. r_{ij} = randomly chosen point in the interval I_{ij}
2. r_{ij} = mid-point of the interval I_{ij}.

We implemented SPE with the above modifications and ran it on synthetic data, generated according to this scheme:

1. Generate a random layout of n random integral coordinates on the positive x-axis.
2. For each pair of generated points p_i and p_j, separated by a distance of r_{ij}, define an asymmetric interval I_{ij} of fixed size, $[r_{ij} - \delta_1, r_{ij} + \delta_2]$ that simulates the actual interval defined by the lower and upper limits in which the distance between p_i and p_j lies.

Below, we show a screenshot of the output for a set of 15 randomly-generated points, lying in an interval of size 50 (Fig. 4).

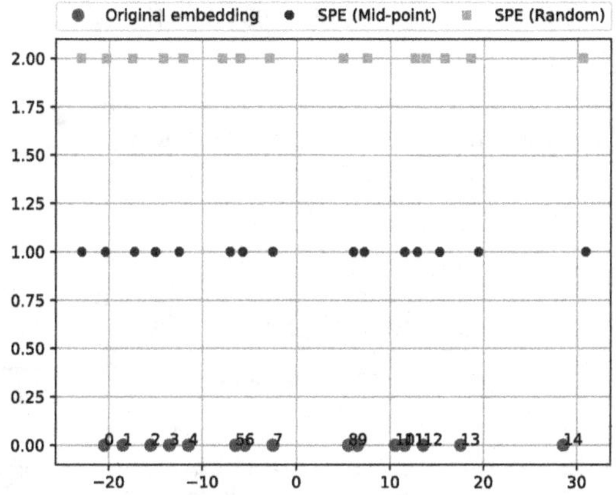

Fig. 4. Fifteen points in an interval of size 50

For a more comprehensive analysis of the behavior of the modified SPE heuristic, we ran our implementation with the two different modifications of line 7, on sets of $10, 20, \ldots, 100$ points, each lying in an interval of size 1000. For each set of points and each modification, we iterated 10 times and computed the average and minimum values of the stress function of Eq. (7) over the 10 iterations. The results are shown in Tables 2 and 3 and pictorially in Figs. 5 and 6 below. We have also compared the run times for these 10 sets of points for both choices of a representative distance, shown in two tables and their respective plots.

Table 2. No. of points vs Stress function (Min)

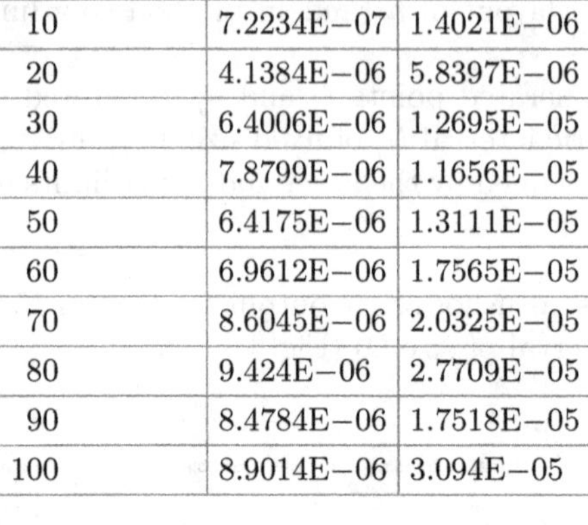

No. of Points	Mid-point	Random
10	7.2234E−07	1.4021E−06
20	4.1384E−06	5.8397E−06
30	6.4006E−06	1.2695E−05
40	7.8799E−06	1.1656E−05
50	6.4175E−06	1.3111E−05
60	6.9612E−06	1.7565E−05
70	8.6045E−06	2.0325E−05
80	9.424E−06	2.7709E−05
90	8.4784E−06	1.7518E−05
100	8.9014E−06	3.094E−05

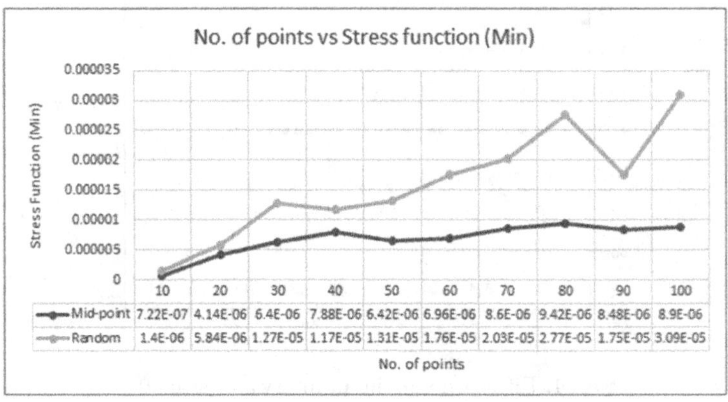

Fig. 5. No. of points vs Stress function (Min)

Table 3. No. of points vs Stress function (Ave)

No. of Points	Mid-point	Random
10	7.1082E−06	1.6575E−05
20	9.1579E−06	0.00015597
30	1.0286E−05	0.00036529
40	1.1951E−05	4.9241E−05
50	9.8893E−06	0.00115544
60	1.3129E−05	0.00914664
70	1.1029E−05	0.00031186
80	0.00604143	0.00819936
90	1.0053E−05	0.00022818
100	1.0678E−05	0.00883982

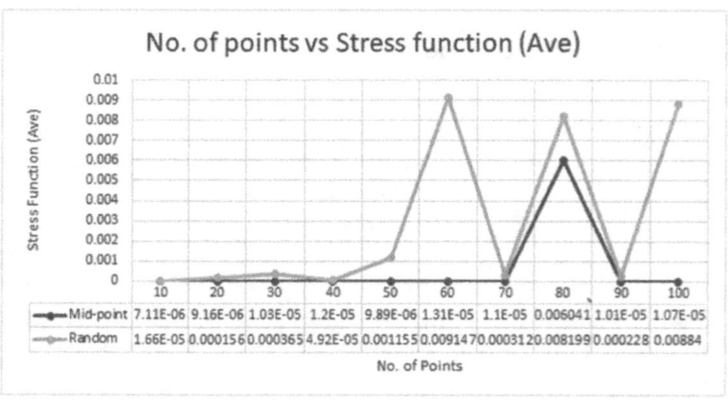

Fig. 6. No. of points vs Stress function (Ave)

Table 4. No. of points vs Time in Secs (Min)

No. of points	Mid-point	Random
10	0.88568419	0.85344799
20	8.68627144	8.76578647
30	34.433548	33.8484532
40	91.3379248	91.3452303
50	200.069031	199.478856
60	375.798189	374.046476
70	628.893871	627.12381
80	1001.19984	1004.2316
90	1584.03128	1563.82617
100	2301.8433	2306.6589

It can be seen both from the stress tables and their corresponding figures that the embedding quality is better when r_{ij} is chosen as the mid-point instead of a random point in a distance interval. The spikes in the number of points versus stress as seen in Fig. 6 is rather intriguing. This is in sharp contrast to the consistency in the times taken when the mid-point or a random point of an interval is selected (Figs. 7, 8 and Tables 4, 5).

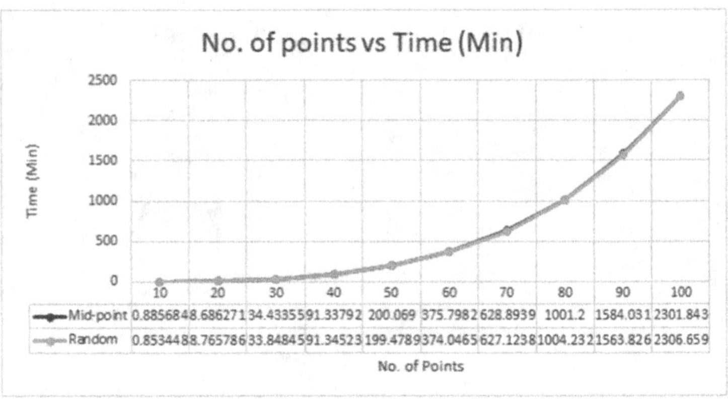

Fig. 7. No. of points vs Time (Min)

Table 5. No. of points vs Time in Secs (Ave)

No. of points	Mid-point	Random
10	0.97215645	0.9172152
20	9.86354464	10.816455
30	37.2580854	36.4575087
40	92.1311842	92.6342256
50	245.667167	238.126872
60	390.279799	385.557976
70	658.691491	643.062603
80	1572.99445	1197.90982
90	1599.38291	1604.16162
100	2432.57755	2452.41778

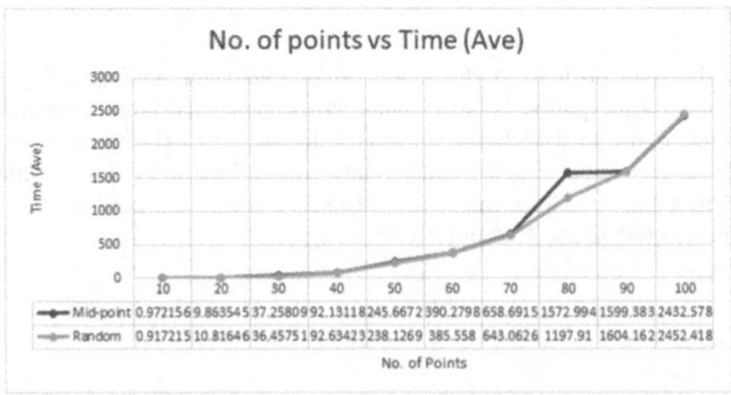

Fig. 8. No. of points vs Time (Ave)

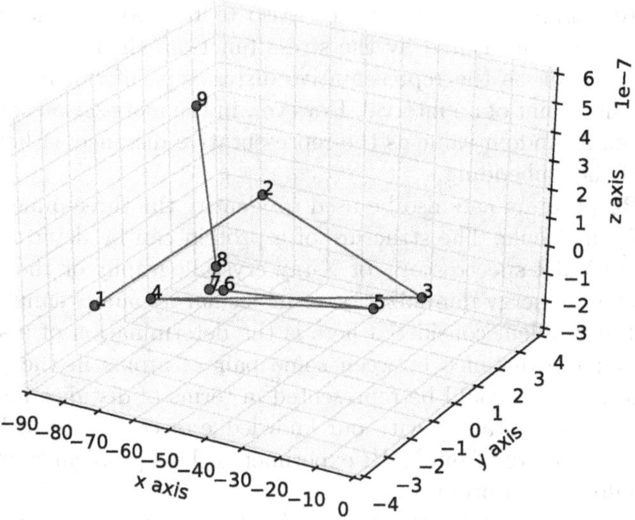

Fig. 9. A sample three-dimensional embedding of 9 points generated by the DGPL program

6 Conclusions

Summary of Work Done: The probe-location problem is an important one in Computational Genomics. In this paper we have proposed an efficient algorithm for its solution. The novelty of our contribution lies in the use of the distance geometry framework. The core of this work took the form of experiments. A representative set of results from these experiments have been presented in an earlier section. Both Mumey's algorithm (see Sect. 3) and the DGPL were tested with different sets of points, ranging in size from 3 to 101, each with a different set of unknown distances. As the results show, the run-time of Mumey's algorithm is highly sensitive to the number of unknown distances, while that of DGPL is almost impervious to this.

We have also explored the viability of using the novel and simple Stochastic Proximity Embedding heuristic to the point placement problem in the inexact model (and hence also the probe location problem). Our extensive experiments show that it is an extremely effective approach to this problem.

Discussions: Of the three approaches, DGPL has far better run-times than the other two methods. It must be noted though that as the final computation of coordinates in the DGPL method depends on an eigenvalue computation it is very sensitive to any perturbations of the data. As can be seen, the SPE method also slows down as we increase the number of points. It is worth pointing out that the iteration loops in the SPE method has to be carefully chosen, making sure that the product $C * S$ is of lower order than n^2, where n is the number of points. Increasing the number of points really takes a toll on the time that the linear programming approach takes. Also to be noted is that the quality of the embedding as measured by the stress function when a random value in an interval is chosen as the representative distance is inferior as compared to choosing the mid-point of an interval. However, in the metrization step of DGPL we have chosen a random value as the representative distance without affecting the quality of the embedding.

The DGPL program can also be used to identify the three-dimensional conformation of a molecule. The structure of a protein can be determined experimentally using NMR spectroscopy or X-ray crystallography or theoretically by means of potential energy minimization or molecular dynamics simulation. More specifically, the problem considered here is the determination of a structure of a protein given the distance between some pair of atoms in the protein and the unknown distances could be represented in terms of distance intervals. The known distances are obtained with our knowledge of certain bond lengths and bond angles or estimated from NMR experiments. This problem is known as the molecular conformation problem.

The steps applied to generate the coordinates in three-dimensions using the DGPL program are slightly different from the generation of the coordinates in one-dimension. The input to DGPL are a set of upper and lower distance intervals for each pair of points, with three as the embedding dimension. Steps 1 to 5 are the same in this case also. Finally, the eigenvalue decomposition of the B matrix is obtained and the product of the three largest eigenvalues with their corresponding normalized eigenvectors yield the three-dimensional coordinates of all the points. A sample plot of a three-dimensional embedding generated by DGPL is shown in Fig. 9.

Future Directions: Further work can be done on several fronts. Except for the point placement problem in one dimension, the challenging problem of finding lower bounds on the number of pairwise distances needed to embed a set of points in three and higher dimensions has not been addressed in the surveyed literature.

In our earlier approaches to the point placement problem, when the basic point-placement graph (such as the 5-cycle, 5:5 jewel etc.) was not rigid, we formulated rigidity conditions that were met over two or more rounds of queries. Now, if the distances returned by the adversary are not exact, there arises the challenging problem of satisfying the rigidity conditions, using exact lengths as well as the distance bounds returned by the adversary, over multiple query rounds.

References

1. Blumenthal, L.M.: Theory and Applications of Distance Geometry (1970)
2. Chin, F.Y.L., Leung, H.C.M., Sung, W.K., Yiu, S.M.: The point placement problem on a line – improved bounds for pairwise distance queries. In: Giancarlo, R., Hannenhalli, S. (eds.) WABI 2007. LNCS, vol. 4645, pp. 372–382. Springer, Heidelberg (2007). https://doi.org/10.1007/978-3-540-74126-8_35

3. Damaschke, P.: Point placement on the line by distance data. Discrete Appl. Math. **127**(1), 53–62 (2003)
4. Malliavin, T.E., Mucherino, A., Nilges, M.: Distance geometry in structural biology: new perspectives. In: Mucherino, A., Lavor, C., Liberti, L., Maculan, N. (eds.) Distance Geometry, pp. 329–350. Springer, New York (2013). https://doi.org/10.1007/978-1-4614-5128-0_16
5. Mukhopadhyay, A., Sarker, P.K., Kannan, K.K.V.: Randomized versus deterministic point placement algorithms: an experimental study. In: Gervasi, O., et al. (eds.) ICCSA 2015. LNCS, vol. 9156, pp. 185–196. Springer, Cham (2015). https://doi.org/10.1007/978-3-319-21407-8_14
6. Crippen, G.M., Havel, T.F.: Distance Geometry and Molecular Conformation, vol. 74. Research Studies Press, Somerset, England (1988)
7. Mumey, B.: Probe location in the presence of errors: a problem from dna mapping. Discrete Appl. Math. **104**(1), 187–201 (2000)
8. Agrafiotis, D.K.: Stochastic proximity embedding. J. Comput. Chem. **24**(10), 1215–1221 (2003)
9. Skiena, S.S., Smith, W.D., Lemke, P.: Reconstructing sets from interpoint distances (extended abstract). In: SCG 1990: Proceedings of the Sixth Annual Symposium on Computational Geometry, pp. 332–339. ACM, New York (1990)
10. Smith, H., Wilcox, K.W.: A restriction enzyme from hemophilus influenzae. I. Purification and general properties. J. Mol. Biol. **51**, 379–391 (1970)
11. Alam, M.S., Mukhopadhyay, A.: Three paths to point placement. In: Ganguly, S., Krishnamurti, R. (eds.) CALDAM 2015. LNCS, vol. 8959, pp. 33–44. Springer, Cham (2015). https://doi.org/10.1007/978-3-319-14974-5_4
12. Young, G., Householder, A.S.: Discussion of a set of points in terms of their mutual distances. Psychometrika **3**(1), 19–22 (1938)
13. Emiris, I.Z., Psarros, I.D.: Counting Euclidean embeddings of rigid graphs. CoRR abs/1402.1484 (2014)
14. Saxe, J.B.: Embeddability of weighted graphs in k-space is strongly NP-hard. In: 17th Allerton Conference on Communication, Control and Computing, pp. 480–489 (1979)
15. Mukhopadhyay, A., Rao, S.V., Pardeshi, S., Gundlapalli, S.: Linear layouts of weakly triangulated graphs. Discrete Math. Algorithms Appl. **8**(3), 1–21 (2016)
16. Mukhopadhyay, A., Sarker, P.K., Kannan, K.K.: Point placement algorithms: an experimental study. Int. J. Exp. Algorithms **6**(1), 1–13 (2016)
17. Havel, T.F.: Distance geometry: Theory, algorithms, and chemical applications. In: Encyclopedia of Computational Chemistry, vol. 120 (1998)
18. Newell, W.R., Mott, R., Beck, S., Lehrach, H.: Construction of genetic maps using distance geometry. Genomics **30**(1), 59–70 (1995)
19. Dress, A.W.M., Havel, T.F.: Shortest-path problems and molecular conformation. Discrete Appl. Math. **19**(1–3), 129–144 (1988)
20. Cormen, T.H., Leiserson, C.E., Rivest, R.L.: Introduction to Algorithms. The MIT Press and McGraw-Hill Book Company, Cambridge, New York (1989)
21. Buetow, K.H., Chakravarti, A.: Multipoint gene mapping using seriation. I. General methods. Am. J. Hum. Genet. **41**(2), 180 (1987)
22. Pinkerton, B.: Results of a simulated annealing algorithm for fish mapping. Communicated by Dr. Larry Ruzzo, University of Washington (1993)
23. Redstone, J., Ruzzo, W.L.: Algorithms for a simple point placement problem. In: Bongiovanni, G., Petreschi, R., Gambosi, G. (eds.) CIAC 2000. LNCS, vol. 1767, pp. 32–43. Springer, Heidelberg (2000). https://doi.org/10.1007/3-540-46521-9_3

Cinolib: A Generic Programming Header Only C++ Library for Processing Polygonal and Polyhedral Meshes

Marco Livesu(✉)

CNR IMATI, Genoa, Italy
marco.livesu@gmail.com

Abstract. Inspired by the recent growth of computational methods for general polygonal and polyhedral meshes, this paper introduces Cinolib: a novel header only C++ library for geometry processing. Cinolib differentiates itself from similar toolkits in that it is specifically designed to support a wide set of meshes, such as triangle, quadrilateral and general polygonal surface meshes, as well as tetrahedral, hexahedral and general polyhedral volumetric meshes. At the core of the library there is a hierarchical data structure that factorizes the common properties among the various meshes, allowing tools and algorithms to operate on the widest possible set of meshes with a single implementation, thus avoiding code repetition and facilitating bug fixing and software maintenance. Cinolib is licensed with MIT, it currently counts more than 50K lines of code and, besides the core structure, already comprises a vast set of widespread tools for computer graphics and engineering.

Keywords: Mesh processing · Geometry processing · Scientific visualization · Software

1 Introduction

Performing some computation on a geometric domain often requires a discretization of it. In computer graphics and engineering complex geometric domains are typically split into simpler elements seamlessly attached to one another, generating a so called *mesh*. For historical reasons, and because of their nice geometric properties, most of the theory and practical algorithms are tailored for meshes made of canonical elements, that is, triangles and quads for surfaces, and tetrahedra and hexahedra for volumes. However, the generation of meshes that satisfy the minimum quality requirements imposed by the analysis they will undergo is a complex problem, which may take up to 8 times the time necessary to perform the analysis itself [13]. A recent trend in literature tries to extend the analysis to general polygons and polyhedra, thus relaxing the constraints for meshing algorithms, and ultimately simplifying the meshing problem. This is for example the case of the PolySpline method [27] and the Virtual Element Method [36], which

M. L. Gavrilova and C. J. K. Tan (Eds.): Trans. on Comput. Sci. XXXIV, LNCS 11820, pp. 64–76, 2019.
https://doi.org/10.1007/978-3-662-59958-7_4

can be seen as extensions of classical finite element methods to domains containing general polygonal and polyhedral elements. Mesh processing tools offered by the research community do not natively support this growth of computational methods for general meshes: MeshLab [6] and the underlying VCG [15], as well as ImatiSTL [2], are specifically designed for triangle meshes and besides visualization offer little to none support for other surface meshes; libIGL [14] supports both surfaces and volumes, but does not extend to general polygons or polyhedra. Its extension to general polygons, libhedra [1], allows to represent a wider class of meshes but still sacrifices flexibility to maintain a fixed memory layout, and is devoted to surfaces only, without supporting general polyhedral meshes. INRIA's GEOGRAM [16] offers a wider support for volumetric meshes, but the elementary cell elements must still be of a finite set of types (tets, hexa, prisms, cones), and more complex elements can only be achieved by clustering together multiple elementary elements. Finally, OpenVolumeMesh [3] extends to volumes the DCEL approach of OpenMesh [3] and is quite general, but the two libraries remain separated to one another, and do not offer a unified framework for the processing of surfaces and volumes.

This paper presents *Cinolib*: a C++ mesh processing library which aims to support the recent growth of computational methods for general meshes. Differently from previous geometry processing tools, Cinolib is specifically designed for general surface and volumetric meshes, and allows for an intuitive and easy to maintain code-base, where algorithms and operators are implemented once and applied to multiple mesh types, thus avoiding code repetition and facilitating bug fixing and software maintenance. Cinolib is header only, it is licensed with MIT, and is already publicly available on GitHub (https://github.com/mlivesu/cinolib). Using templates to define the attributes associated to each mesh element, it is both highly customizable and straightforward to compile.

2 Software Description

Cinolib's flexibility is made possible by a hierarchical mesh data structure that embraces both surfaces and volumes, and is at the core of the library (Sect. 2.1). Overall, the library comprises around 50K lines of code, which include both the core, and a variety of widespread geometry processing tools to support mesh generation, visualization, analysis, and so forth (Sect. 2.2).

2.1 Architecture

In Cinolib the connectivity of each mesh is represented as a set of elements' lists and their adjacency, which are collocated in a hierarchical data structure that comprises both surfaces and volumes. As depicted in Fig. 1, the hierarchy is a tree. The root summarizes all the properties that are common to *any* mesh, whereas the leaves are the actual meshes the user can create, that is: triangle meshes, quad meshes, general polygon meshes, tetrahedral meshes, hexahedral meshes, and general polyhedral meshes (Fig. 2). In between there is a middle

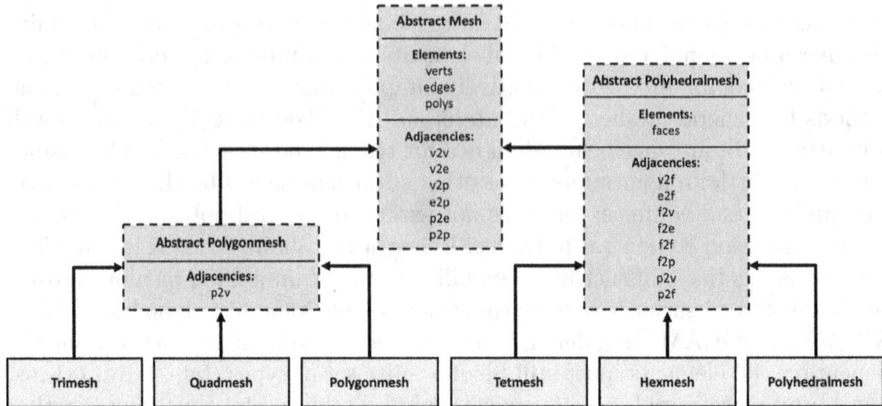

Fig. 1. The mesh hierarchy at the base of Cinolib. Yellow boxes represent the meshes available in the library; gray boxes are the abstract classes from which they inherit the structure. Elements and adjacencies that are common to both surface and volume meshes are stored at the root of the hierarchy (the *p* in AbstractMesh denotes a *polygon* for a surface mesh, and a *polyhedron* for a volume mesh). (Color figure online)

layer, which contains two nodes that summarize common properties which are specific to surface meshes and volumetric meshes, respectively. The philosophy of Cinolib is to make sure that algorithms operate at the highest possible level in the hierarchy, so that they can be applied to the widest set of meshes with a unique implementation. This has clear advantages, as it reduces the coding effort and greatly simplifies bug fixing and software maintenance in general. As a practical example the reader may think of the Dijkstra's algorithm for shortest path computation on a graph. If we consider the *primal* mesh, that is the graph having as nodes the mesh vertices and as arcs the mesh edges, Dijkstra operates indistinctively on surfaces and volumes. In Cinolib Dijkstra's routines operate at the root of the hierarchy, thus the same piece of code is used by all the meshes supported in the library. Let us now consider the *dual* mesh, that is the graph having one node per element, and one arc for each pair of adjacent elements. Here there is a difference between surfaces and volumes, as elements are polygons in the former, and polyhedra in the latter. To maximize compatibility between surface and volume meshes a strict naming convention is used throughout the whole library. Surface meshes are defined as lists of: verts (v), edges (e) and polys (p); volume meshes are defined as lists of: verts (v), edges (e), faces (f) and polys (p). The word polys appears both in surfaces and volumes, but denotes polygons in the former and polyhedra in the latter. This double meaning of the term polys is exploited in the hierarchy, which defines polys at the top level and allows algorithms that work on the dual mesh and only require poly to poly adjacency (p2p) to operate indistinctively on a surface or volumetric mesh. This is the case of Dijkstra's on the dual mesh, but also of methods to compute spanning trees and various clustering algorithms that start associating a label to one element

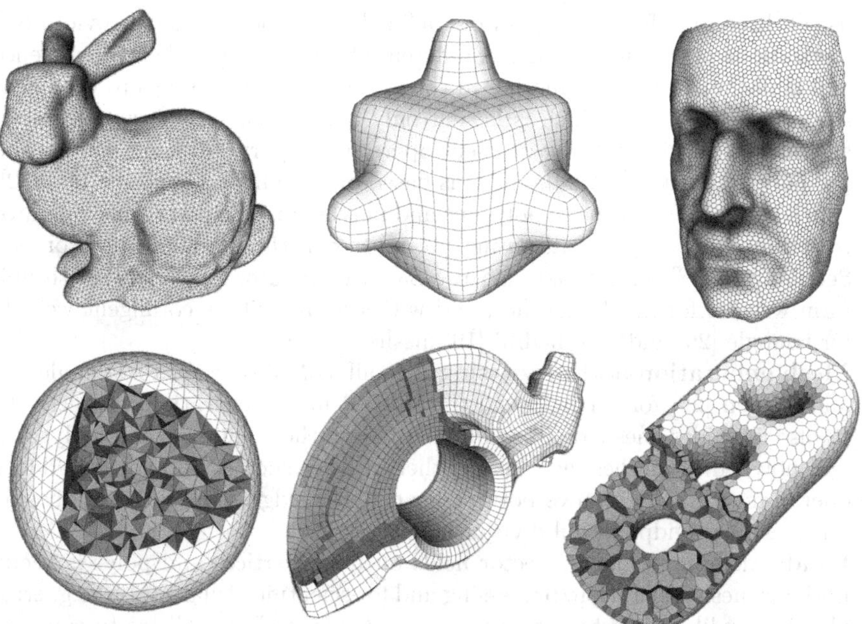

Fig. 2. From left to right, top to bottom: triangle mesh, quadmesh, general polygon mesh, tetmesh, hexmesh, general polyhedral mesh. All these meshes are supported by Cinolib.

and expand the cluster by conquering adjacent elements. All these algorithms are implemented once, and used by all the surface and volumetric meshes in the library. Similarly, all methods that apply only to surfaces (or volumes) but are not specific to a particular mesh type, are implemented at the middle level of the hierarchy, and the same code covers all the surface (or volumetric) meshes supported in the library. This is the case of various operators that allow to edit the mesh connectivity (e.g. addition/removal of mesh elements), computation of gradients and iso contours of functions encoded at mesh vertices, and so forth.

2.2 Functionalities

Besides the mesh hierarchy, Cinolib offers a variety of tools and algorithms that cover a wide specturm of ubiquitous operations in computer graphics and engineering. This section provides a non exhaustive lists of its key features. Unless stated differently, all the features are implemented directly in the code base and do not depend from external software, making the library extremely easy to compile and use

- **Discrete Differential operators:** many tools in computer graphics and engineering require solving Poisson and Laplace problems, and make extensive use of discrete differential operators defined on meshes. Cinolib provides discretizations of the most common differential operators (gradient, divergence and laplacian) for all the meshes in the library, as well as wraps to the linear solvers of Eigen [12], which is header only and therefore as easy to compile as Cinolib. Specifically, the gradient operator supports both per vertex and per poly gradient fields [24], and it is based on the Green-Gauss method [32], which works on any type of polygons and polyhedra. The divergence operator is obtained by simply transposing the matrix of the gradient operator (see Section 3 in [18] for details), and the Laplace operator is available with uniform weights for all the meshes, and with the ubiquitous cotangent weights for triangle [26] and tetrahedral [19] meshes;

- **Mesh generation and processing:** Cinolib contains wraps to Triangle [28] and Tetgen [29] for triangle and tetrahedral mesh generation, respectively. It also provides mesh dualization to convert them into general polygonal and polyhedral meshes, as well as other utilities such as topological editing operators (e.g. add/remove polys, edge collapse, edge split), and subdivision schemes (e.g. midpoint subdivision);
- **Fields:** defining scalar or vector fields on mesh vertices and polys is useful both for mesh generation/processing and for scientific visualization (e.g. error plots). Cinolib supports per vertex scalar fields, and also allows to compute and visualize iso-lines of fields embedded on surface meshes, or iso-surfaces of fields embedded on volumetric meshes. Thanks to the aforementioned topological editing operators both iso-lines and iso-surfaces can be optionally embedded in the mesh connectivity, splitting the edges they traverse (Fig. 3). Vector fields are also supported, as well as tools to process and visualize integral lines that align to them (Fig. 4);

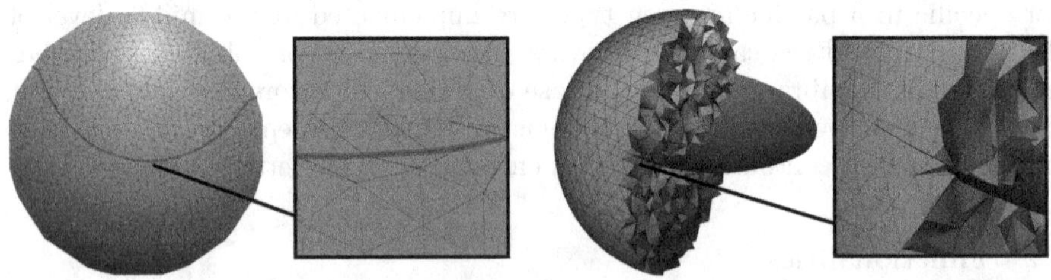

Fig. 3. Level sets of a scalar field embedded on the vertices of a surface mesh (left) and volumetric mesh (right). Curves and surfaces can be optionally embedded in the mesh connectivity by splitting the edges they traverse (closeups). This latter operation is currently supported only for triangle and tetrahedral meshes.

- **Distances and paths:** Cinolib contains various implementations of Breadth-First Search (BFS) and shortest paths computation (using Dijkstra [9]). Both algorithms can run on both the primal and the dual mesh, support optional

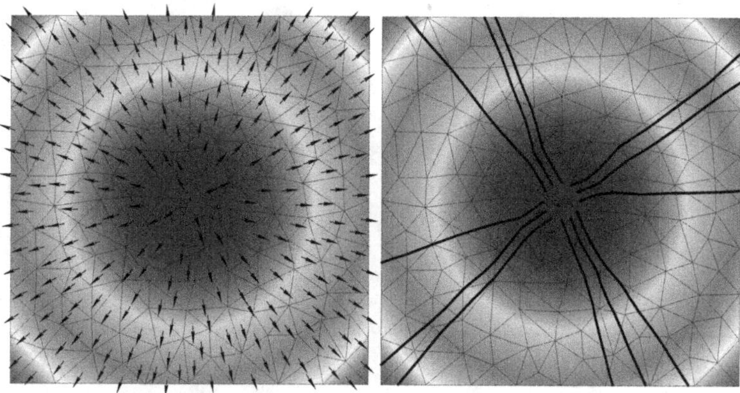

Fig. 4. Left: a scalar function embedded on the vertices of a surface mesh, and its (per poly) gradient field. Right: four bundles of integral lines that emanate from the center and align to the gradient field. The figure is taken from [24].

constraints (e.g. barriers), and rely on a unique implementation for all the meshes in the hierarchy. The library also contains an implementation of the heat-based geodesics [7], which also operates at the root of the mesh hierarchy and can be applied to any mesh in the library. Heat geodesics are computed solving an initial value problem rather than a boundary problem. This allows to factorize the matrix of the heat flow operator once, and then solve the geodesic problem as many times the user wishes in real time by means of a simple back-substitution (less than 0.005 s on the bunny shown in Fig. 5, which contains 14K vertices);

- **Visualization/rendering:** various tools for rendering and visualization are offered to the user, spanning from triangulation of arbitrary polygons to plot general polygonal/polyhedral meshes, to various colors schemes and 1D/2D texture facilities. Popular color maps such as the HSV ramp (Fig. 4) and Parula (Fig. 3), as well as textures commonly used to show error plots and distortion maps in scientific papers (Fig. 6) are also included in the library and are available for all the meshes in the hierarchy. Additionally, Cinolib provides an OpenGL canvas with trackball and perspective/orthograpic camera, and includes a slicing tool that allows to cut any mesh with axis aligned planes and show/hide only a portion of it (Fig. 2). The computation of ambient occlusion [25] for realistic and more revealing rendering of complex 3D scenes is also supported (Fig. 7);
- **3D printing:** facilities for additive manufacturing are also included in the library. Specifically, Cinolib allows to read a set of slices and transform them into a polygon mesh that can be used for both visualization and analysis. Since support structures are typically encoded as 1D lines and their ultimate shape depends on the material and hardware used for print [20], a tool for their thickening and merging is included, so that the user can set the proper

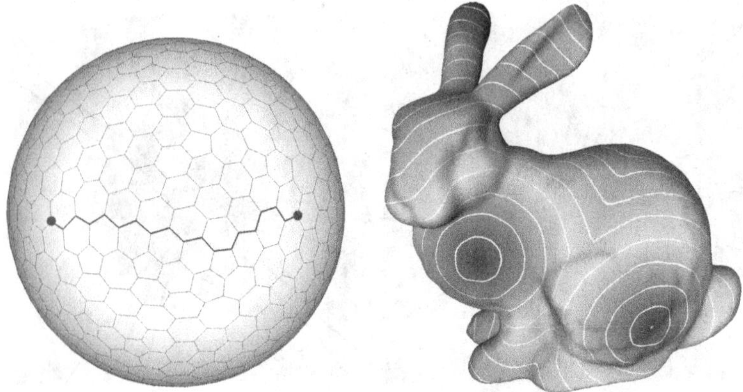

Fig. 5. Left: shortest path between two mesh vertices of a general polygonal mesh, computed with Dijkstra. Right: heat-based geodesics computed with [7].

thickening radius and check how supports will look like in the actual object before the print happens (Fig. 8);

- **IO:** Cinolib currently supports a wide set of popular file formats in the computer graphics and engineering community, such as OFF, OBJ and IV for surfaces, and MESH, VTU, VTK, TET for volumes. Since general polyhedral meshes are not well supported by these formats, we also added IO facilities for the HYBRID format defined by the authors of [11], as well as proposed a new format (HEDRA) for such meshes. Additionally, Cinolib can read CLI files for 3D printing facilities, and can read or write curve skeletons encoded in various ad-hoc file formats used by the authors of relevant papers in the field, such as [8,21,23,34];
- **Utilities:** additional utilities are also available, such as wraps for the min-cut and graph-cut algorithms [4]; profiling utilities to measure timings and improve performances; quality metrics for polygons (e.g. maximum inscribed circle, minimum outer circle, kernel), tetrahedra and hexahedra (implementing [33]); computation of coarse layouts for quad and hex meshes [5,30]; tools for the analysis and processing of curve-skeletons (used in [19,22,35]), as well as other features not included in this list due to space limits.

3 Illustrative Example

As discussed in Sect. 2.2, Cinolib provides an extensive set of tools and functionalities. Illustrating all of them is out of the scope of the article. This section proposes just a simple code sample that shows how to write a program to compute a harmonic field on a general polygonal mesh. Harmonic functions are at the base of many remeshing and parameterization techniques in computer graphics [10,19]. Given a set of vertices where the function reaches its extrema, computing a harmonic field f amounts to solving a Laplace equation $\Delta f = 0$, subject

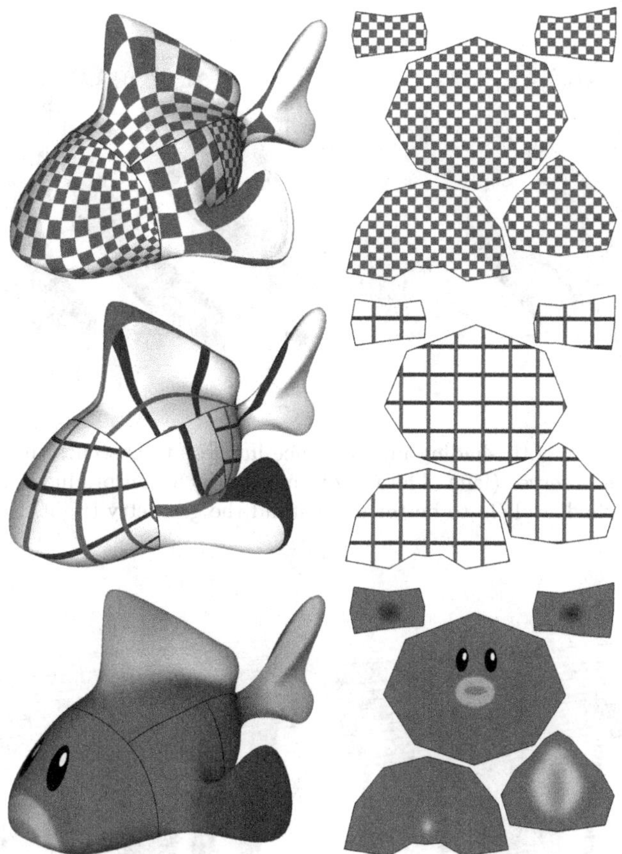

Fig. 6. Cinolib supports popular 2D textures to show distortion of UV maps, such as checkerboards (top) and iso-lines (middle), but also textures loaded from external image files (bottom).

to Dirichlet boundary conditions on the (known a priori) function extrema. Cinolib supports the computation of harmonic fields, as well as their visualization and inspection, with tools such as color maps, iso contours, gradients and integral lines. In the following a simple program for the field computation is shown:

```
#include <cinolib/gui/qt/glcanvas.h>
#include <cinolib/meshes/meshes.h>
#include <cinolib/harmonic_map.h>

void main()
{
    // load a general polygon mesh from file
    cinolib::DrawablePolygonmesh<> m("./bunny.obj");
```

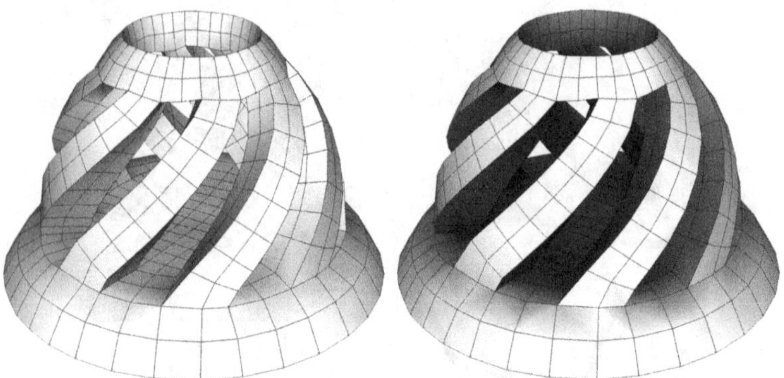

Fig. 7. Standard smooth shading may produce images of 3D scenes that are difficult to parse for the observer (left). The use of ambient occlusion produces more realistic images, where shadows help to better understand the geometry (right)

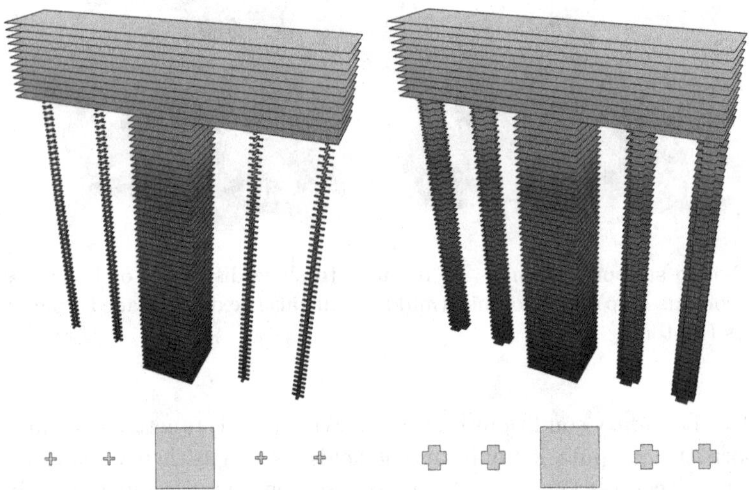

Fig. 8. Two versions of a sliced T-like shape sustained by four columns of supports. External supports are typically encoded as piece-wise linear curves, and their ultimate size and shape depends on the material and hardware used for the 3d print (e.g. the size of the filament in FDM, or the laser beam in SLM/SLS). Cinolib offers functionalities for loading and converting sliced data into polygon meshes that can be visually inspected, as well as thickening and merging of support structures, to check how they will look like in the final print. This figure is taken from [20], where these functionalities were used to convert sliced data into tetrahedral meshes for the simulation of 3d printing processes.

```
    // setup the Dirichlet boundary conditions of the field. We
    // are looking for a function that evaluates 0 at vertex v0
    // and 1 at vertex v100
    std::map<int,double> bc;
    bc[0] = 0.0;
    bc[0] = 1.0;

    // compute a harmonic field f. The Laplace operator is discretized with
    uniform weights cinolib::ScalarField f = harmonic_map(m, bc, 1, UNIFORM);

    // copy the field on the mesh (for visualization)
    f.copy_to_mesh(m);

    // create and show an OpenGL canvas showing the mesh and the field
    cinolib::GLcanvas gui;
    gui.push_obj(&m);
    gui.show();
}
```

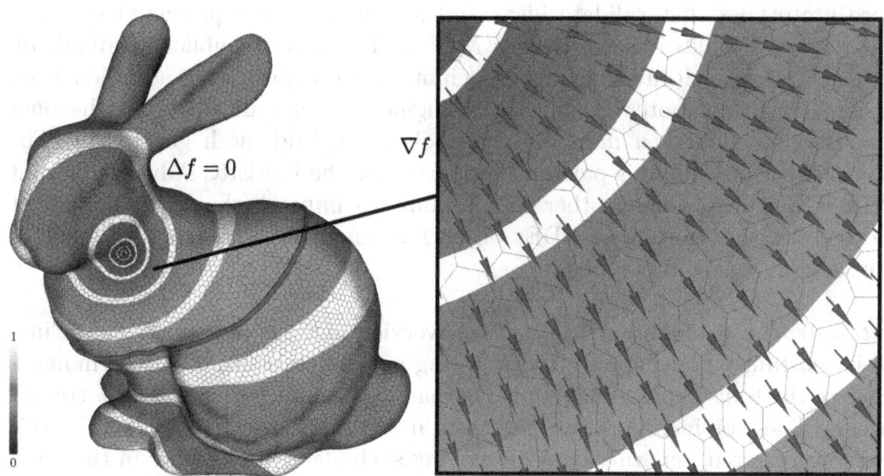

Fig. 9. Example of the solution of a Laplace problem to compute a harmonic function f on a general polygonal mesh of the Stanford Bunny. Colorization is given by the popular Parula colormap, enriched with white bands to show the level sets of the function. In thick red a iso-curve of the field is shown. The closeup shows details of the tessellation and the (normalized) gradient ∇f, computed with the Green-Gauss method, which applies to any polygon and polyhedron [32]. Right: the code used to compute the field and draw it on the canvas. The *harmonic map* routine operates at the highest level of the mesh hierarchy; substituting *DrawableTrimesh* with a different type the very same code could be used to compute similar fields on any other mesh supported by Cinolib, without changing a single line of code. (Color figure online)

As can be noticed from the lines above, the code is extremely concise yet intuitive. Figure 9 shows the result it produces. The routine for computation of harmonic fields, as well as many others based on discrete differential geometry operators, work at the highest level of the mesh hierarchy. Therefore, a single implementation can be re-used on many meshes, without the need to change a single line of code, if not the one where the mesh is loaded. This feature was used to produce all the results shown in [18].

4 Conclusions and Future Works

A new mesh processing C++ library is presented. Compared to alternative libraries, Cinolib is specifically designed to handle surface and volumetric meshes made of general polygons and polyhedra. Its hierarchical mesh data structure allows to easily develop algorithms that naturally scale to multiple mesh types, without repetitions of code occurring. Although it tries to be as efficient as possible, whenever generality and efficiency were in conflict, generality was always pursued. All in all, Cinolib is made by researchers for researchers, and is specifically designed to promote code re-usability and to easily create software prototypes that validate ideas and algorithms to be presented in scientific papers. Cinolib is licensed with MIT, and is already publicly available on GitHub (https://github.com/mlivesu/cinolib). We expect wide adoption from researchers in computer graphics and engineering, specifically from the ones that develop numerical methods and tools for hybrid mesh generation. The development of Cinolib is partially supported by the EU ERC Advanced Grant CHANGE which, among other things, aims to improve the Virtual Element Method for the solution of PDEs on surface and volumetric meshes made of general polygons and polyhedra.

Future Works: while this article gives an overview of the library as of today, Cinolib is constantly growing and incorporating new functionalities. At the moment some of the tools exposed are supported only by simplicial meshes (e.g. tracing integral lines, or incorporating level sets into the mesh connectivity). To this end, future extensions will focus on moving such algorithms higher in the mesh hierarchy in order to extend such functionalities to all the meshes. Regarding new tools to be implemented, priority will be given to: the improvement of connectivity editing operators for general meshes (after [11]) and for quad and hex meshing (e.g. sheet/chord collapse); the implementation of a QP solver and of Lagrange multipliers; the implementation of tools to create and process surface cross field and volumetric frame fields; the implementation of well established methods for mesh deformation [31]; and the introduction of tools for cage- and skeleton-based animation.

Acknowledgements. This work is partially supported by the EU ERC Advanced Grant CHANGE, grant agreement No.694515. We thank Keenan Crane for releasing the fish model in Fig. 6 to the public domain.

References

1. Vaxman, A., et al.: libhedra: geometric processing and optimization of polygonal meshes (2017). https://github.com/avaxman/libhedra
2. Attene, M.: ImatiSTL - fast and reliable mesh processing with a hybrid kernel. In: Gavrilova, M.L., Tan, C.J.K. (eds.) Transactions on Computational Science XXIX. LNCS, vol. 10220, pp. 86–96. Springer, Heidelberg (2017). https://doi.org/10.1007/978-3-662-54563-8_5
3. Botsch, M., Steinberg, S., Bischoff, S., Kobbelt, L.: OpenMesh-a generic and efficient polygon mesh data structure (2002)
4. Boykov, Y., Veksler, O., Zabih, R.: Fast approximate energy minimization via graph cuts. IEEE Trans. Pattern Anal. Mach. Intell. **23**(11), 1222–1239 (2001)
5. Cherchi, G., Livesu, M., Scateni, R.: Polycube simplification for coarse layouts of surfaces and volumes. Comput. Graph. Forum **35**(5), 11–20 (2016). https://doi.org/10.1111/cgf.12959
6. Cignoni, P., Callieri, M., Corsini, M., Dellepiane, M., Ganovelli, F., Ranzuglia, G.: MeshLAB: an open-source mesh processing tool. In: Eurographics Italian Chapter Conference 2008, pp. 129–136 (2008)
7. Crane, K., Weischedel, C., Wardetzky, M.: Geodesics in heat: a new approach to computing distance based on heat flow. ACM Trans. Graph. (TOG) **32**(5), 152 (2013)
8. Dey, T.K., Sun, J.: Defining and computing curve-skeletons with medial geodesic function. In: Proceedings of the Fourth Eurographics Symposium on Geometry Processing, pp. 143–152. Eurographics Association (2006)
9. Dijkstra, E.W.: A note on two problems in connexion with graphs. Numer. Math. **1**(1), 269–271 (1959)
10. Eck, M., DeRose, T., Duchamp, T., Hoppe, H., Lounsbery, M., Stuetzle, W.: Multiresolution analysis of arbitrary meshes. In: Proceedings of the 22nd Annual Conference on Computer Graphics and Interactive Techniques, pp. 173–182. ACM (1995)
11. Gao, X., Jakob, W., Tarini, M., Panozzo, D.: Robust hex-dominant mesh generation using field-guided polyhedral agglomeration. ACM Trans. Graph. (TOG) **36**(4), 114 (2017)
12. Guennebaud, G., Jacob, B., et al.: Eigen v3 (2010). http://eigen.tuxfamily.org
13. Hughes, T.J., Cottrell, J.A., Bazilevs, Y.: Isogeometric analysis: CAD, finite elements, NURBS, exact geometry and mesh refinement. Comput. Methods Appl. Mech. Eng. **194**(39–41), 4135–4195 (2005)
14. Jacobson, A., Panozzo, D., et al.: libigl: a simple C++ geometry processing library (2018). http://libigl.github.io/libigl/
15. CIVC Lab: VCG: Visualization and computer graphics library (2004). https://github.com/cnr-isti-vclab/vcglib
16. Levy, B.: Geogram (2015)
17. Lévy, B., Petitjean, S., Ray, N., Maillot, J.: Least squares conformal maps for automatic texture atlas generation. ACM Trans. Graph. (TOG) **21**, 362–371 (2002)
18. Livesu, M.: A heat flow relaxation scheme for n dimensional discrete hyper surfaces. Comput. Graph. **71**, 124–131 (2018). https://doi.org/10.1016/j.cag.2018.01.004
19. Livesu, M., Attene, M., Patane, G., Spagnuolo, M.: Explicit cylindrical maps for general tubular shapes. Comput.-Aided Des. **90**, 27–36 (2017). https://doi.org/10.1016/j.cad.2017.05.002. SI: SPM2017

20. Livesu, M., Cabiddu, D., Attene, M.: Slice2Mesh: a meshing tool for the simulation of additive manufacturing processes. Comput. Graph. **80**, 73–84 (2019). https://doi.org/10.1016/j.cag.2019.03.004. ISSN 0097-8493
21. Livesu, M., Guggeri, F., Scateni, R.: Reconstructing the curve-skeletons of 3D shapes using the visual hull. IEEE Trans. Visual Comput. Graph. **18**(11), 1891–1901 (2012). https://doi.org/10.1109/TVCG.2012.71
22. Livesu, M., Muntoni, A., Puppo, E., Scateni, R.: Skeleton-driven adaptive hexahedral meshing of tubular shapes. Comput. Graph. Forum **35**(7), 237–246 (2016). https://doi.org/10.1111/cgf.13021
23. Livesu, M., Scateni, R.: Extracting curve-skeletons from digital shapes using occluding contours. Vis. Comput. **29**(9), 907–916 (2013). https://doi.org/10.1007/s00371-013-0855-8
24. Mancinelli, C., Livesu, M., Puppo, E.: A comparison of methods for gradient field estimation on simplicial meshes. Comput. Graph. **80**, 37–50 (2019). https://doi.org/10.1016/j.cag.2019.03.005. ISSN 0097-8493
25. Méndez-Feliu, À., Sbert, M.: From obscurances to ambient occlusion: a survey. Vis. Comput. **25**(2), 181–196 (2009)
26. Meyer, M., Desbrun, M., Schröder, P., Barr, A.H.: Discrete differential-geometry operators for triangulated 2-manifolds. In: Hege, H.C., Polthier, K. (eds.) Visualization and Mathematics III. MATHVISUAL, pp. 35–57. Springer, Heidelberg (2003). https://doi.org/10.1007/978-3-662-05105-4_2
27. Schneider, T., Dumas, J., Gao, X., Botsch, M., Panozzo, D., Zorin, D.: Poly-spline finite element method. CoRR abs/1804.03245 (2018), http://arxiv.org/abs/1804.03245
28. Shewchuk, J.R.: Triangle: engineering a 2D quality mesh generator and delaunay triangulator. In: Lin, M.C., Manocha, D. (eds.) WACG 1996. LNCS, vol. 1148, pp. 203–222. Springer, Heidelberg (1996). https://doi.org/10.1007/BFb0014497
29. Si, H.: TetGen, a delaunay-based quality tetrahedral mesh generator. ACM Trans. Math. Softw. (TOMS) **41**(2), 11 (2015)
30. Sorgente, T., Biasotti, S., Livesu, M., Spagnuolo, M.: Topology-driven shape chartification. Comput.-Aided Geom. Des. **65**, 13–28 (2018). https://doi.org/10.1016/j.cagd.2018.07.001
31. Sorkine, O., Alexa, M.: As-rigid-as-possible surface modeling. In: Proceedings of the Fifth Eurographics Symposium on Geometry Processing, pp. 109–116. Eurographics Association (2007)
32. Sozer, E., Brehm, C., Kiris, C.C.: Gradient calculation methods on arbitrary polyhedral unstructured meshes for cell-centered CFD solvers. In: 52nd Aerospace Sciences Meeting, p. 1440 (2014)
33. Stimpson, C., Ernst, C., Knupp, P., Pébay, P., Thompson, D.: The verdict library reference manual. Sandia National Laboratories Technical report 9 (2007)
34. Tagliasacchi, A., Alhashim, I., Olson, M., Zhang, H.: Mean curvature skeletons. Comput. Graph. Forum **31**, 1735–1744 (2012)
35. Usai, F., Livesu, M., Puppo, E., Tarini, M., Scateni, R.: Extraction of the quad layout of a triangle mesh guided by its curve skeleton. ACM Trans. Graph. **35**(1), 6:1–6:13 (2015). https://doi.org/10.1145/2809785
36. da Veiga, L.B., Brezzi, F., Marini, L.D., Russo, A.: The hitchhiker's guide to the virtual element method. Math. Model. Methods Appl. Sci. **24**(08), 1541–1573 (2014)

Trust Computation in VANET Cloud

Brijesh Kumar Chaurasia[1(✉)] and Kapil Sharma[2]

[1] Indian Institute of Information Technology, Lucknow, UP, India
brijesh@iiitl.ac.in

[2] ITM University Gwalior, Gwalior, MP, India
kapil.sharma.cse@itmuniversity.ac.in

Abstract. In this paper, we present a mechanism for trust computation in VANET cloud. The presence of VANET cloud allows vehicles to store their past trust values of other vehicles. These values are utilized for fast trust computation by other vehicles. The mechanism takes into consideration the uncertainty and fuzziness associate with trust values by incorporating DST (Dempster Shafer Theory) and fuzzy analyzer. The change in trust value due to action of a vehicle is done through a reward & penalty scheme. This proposed mechanism optimizes the execution and computation of a batch of simulations, increasing the overall performance, in terms of simulation time and costs. Simulation results indicate the DST based fuzzy trust mechanism is able to manage the trust of vehicles efficiently in the VANET cloud environment.

Keywords: Cloud computing · CloudAnalyst · TOEFVC · VANET · Trust computation · Theory of evidence · DST · Fuzzy analyzer · VANETsim · Rewards and penalty

1 Introduction

VANET envisage supporting services for intelligent transportation systems (ITS), as collective monitoring of traffic, collision avoidance, vehicle navigation, control of traffic lights, and traffic congestion management by signalling to drivers [1]. VANET is a network of vehicles, road side infrastructure and a trusted authority. It is a branch of mobile ad hoc network (MANET) [2]. VANET has many prominent characteristics from other ad hoc networks i.e. vehicles moving at high speeds on the road, dynamic topology, predictable and short interaction time between nodes with limited transmission range. In VANET, vehicles can communicate with nearby vehicles known as a vehicle-to-vehicle (V2V) communication and also with road side infrastructure also known as vehicle-to-infrastructure (V2I) [3]. The primary application of VANET is the dispersion of information about local traffic or road conditions. VANET support two types of messages first are safety message and another, non safety message. Safety messages have high priority than non safety messages. In VANET, the safety related messages broadcast by vehicles are used by other vehicles to make critical decisions. Authenticated vehicles may also send wrong messages. This may trap a vehicle taking wrong decision with dire consequences. So, filtration of correct information from the broadcast messages is the main issue in VANET [4]. Trust management of unauthenticated or malicious vehicles as

M. L. Gavrilova and C. J. K. Tan (Eds.): Trans. on Comput. Sci. XXXIV, LNCS 11820, pp. 77–95, 2019.
https://doi.org/10.1007/978-3-662-59958-7_5

well as authenticated or good vehicles is equally required in VANET. The main application of VANET is the dissemination of information about traffic, road conditions and infotainment and safety critical information but, the issue is how to know the correctness of message sent by another vehicle. Trust establishment between vehicles is, therefore, a major challenge to be solved to ensure the correctness of messages [5]. Trust is the level of confidence on messages sent by a vehicle reporting an event [6]. Reputation is the opinion of one vehicle about other vehicle in VANETs. The vehicle maintains the reputation of other vehicles and uses it to evaluate their trustworthiness [7]. Trust in VANET may be direct trust and indirect trust. Direct trust rapport build from direct communication with some other vehicle within transmission range of source vehicle i.e. vehicle A has a direct trust on vehicle B if they have direct communication. Indirect trust can be formed from reference of other trusted vehicles [8]. Indirect trust degree can be calculated by taking suggestions from most similar nearest neighbour's vehicles as a suggestion trust degree [9]. Internet access in VANET is provided by road side infrastructure [10]. Due to high mobility and rapidly changing network topology connectivity is an issue of VANET. Another problem is that verification of huge message and correctness within time is also a big challenge of VANET. Cloud computing is a promising low latency and context awareness computing paradigm.

Cloud computing [11] refers to the supplying not only computational resources, but services on demand such that infrastructure as a service, storage as a service and platform as a service that means service on demand. In addition, cloud computing could save considerable time since businesses will not bear to install and/or upgrade applications. It is essential to continuously connect to the Internet for stably ubiquitously obtaining adaptable cloud services on the road. In this work, we focus on the trust computation using the theory of evidence with the fuzzy analyzer in VANET cloud (TOEFVC) trust management scheme to minimize the computation time and maximise the accuracy. Our main aim is to solve the issues or problem of trust computing with the help of proposed TOEFVC model within the given time constraints. The rest of this report is organized as follows. Related work is discussed in Sect. 2. The problem formulation is given in Sect. 3. Proposed architecture of TOEFVC is shown in Sect. 4. Simulation and results are discussed in Sect. 5. And finally, Sect. 6 briefly concludes the contribution.

2 Related Work

In existing literature very few cloud based trust computation schemes is available. Shifting VANET to cloud is presented in [12]. In this work, discovering and consuming services within vehicular clouds or CROWN cloud service discovery protocol for VANET and mobile cloud server the name of transportation server, or STAR is proposed. To provide a cloud services an RSU will store each STAR. However, security issue for the proposed system is future scope. The mobile vehicular cloud research issues, design principles and review cloud applications ranging from urban sensing to intelligent transportation is presented in [13]. The work is also identified the role of V2V communications for the propagation of data to facilitate its search (in urban sensing) and for the support of distributed processing (in local route optimization).

Taxonomy of vehicular cloud computing (VCC) is briefly introduced in [14]. The merits of VCC in terms of data aggregation using cloud storage are elaborated in detailed. Cloud storage may be used by private and government department to store certificates, keys and many important security credentials. The VCC can be used as several services such as network as a service (NaaS) to provide moving cars on the road internet facility through mobile phones and fixed access points. Cooperation as a service (CaaS) in VANET provides traffic information (warnings of traffic jams and accidents, weather or road conditions), parking availability and advertisements and driver safety information. Data dissemination in VANET using CaaS is presented in [15]. Cluster based approach is used in CaaS services. In this approach, tree represents a cluster where the root of the tree (i.e. clusterhead). In addition with, CBR is used for intra-cluster communication and DTN routing for inter-cluster communication. Storage as a service (STaaS) provides the vehicles with additional storage capability. To mitigate data corruption, replay and pollution attacks secure and efficient remote data checking for network coding-based distributed storage systems that rely on un-trusted servers is discussed in [16]. Computing as a service (CaaS) provides parking; toll based services [17] and also assists the reveals of parking credentials as parking hours and parking lot etc. [14]. We have applied fuzzy based mechanism in VANET cloud. However, very few fuzzy based trust computation schemes are available in literature. The trust uses fuzzy logic to allow matches to represent in [18, 19]. The system is flexible and fuzzy rules are determined by a scheme designer. Peers are able to scale inputs accordingly. Subjective logic must not be confused with fuzzy logic. The latter operates on crisp and certain measures about linguistically vague and fuzzy propositions whereas subjective logic operates on uncertain measures about crisp propositions [20]. However, secure VANET cloud computing trust management scheme is needed.

3 Problem Formulation

The main requirement of VANET trust computation mechanism is that the formation of trust must take very small time especially in case of safety messages. Hence, VANET is high density network due to verification of each messages may overloaded the network. Trust evaluation of messages at the time of driving is required for safety journey. Other problems are the need to dynamically updated trust values of vehicles, handling uncertain evidence and subjectivity in the interpretation of the same information by different vehicles. The indecision leads to uncertainty that is different from randomness. The technique must also be capable of handling non exact values. Hence, a technique which can cope up with uncertainty is required.

4 Proposed Methodology

In this section, we present a theory of evidence with the fuzzy analyzer in VANET cloud (TOEFVC) to heighten the security of vehicles and information retrieval within time. The problem associated with paucity of inter-vehicular interaction, rapid change

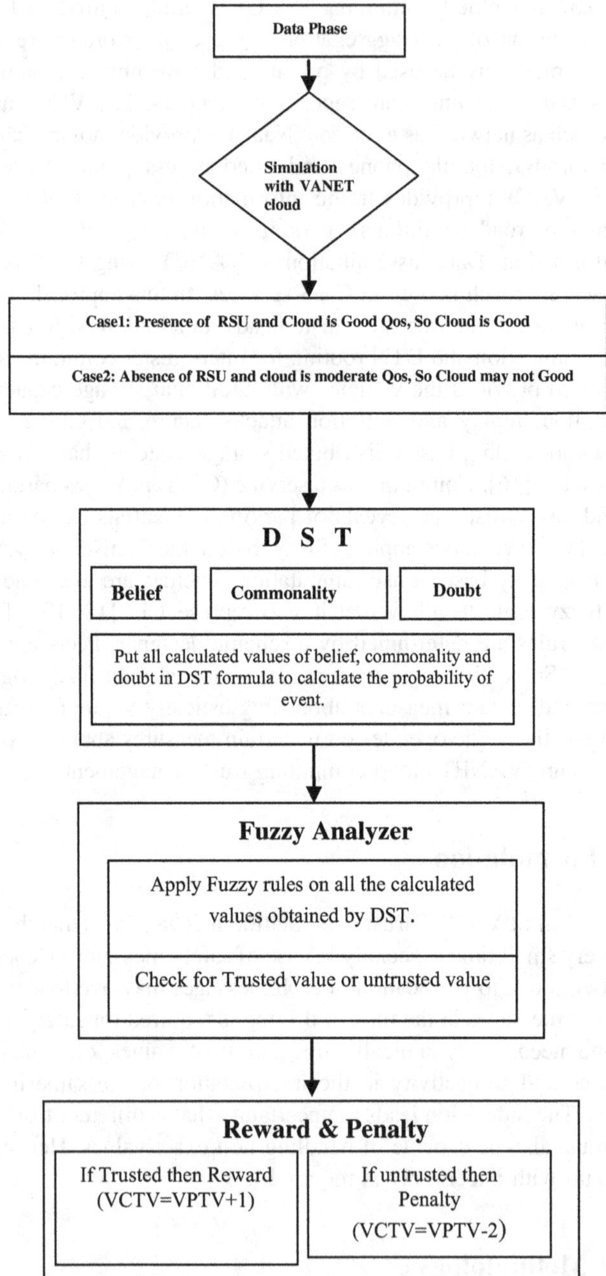

Fig. 1. Proposed TOEFVC model

of neighborhood, information uncertainty and need for continuous modification of trust values can be solved by using a combination of the following. A priori trust data can be obtained from the VANET cloud. The cloud can be queried by a vehicle to get trust information and behavior trajectories of other vehicles. Lack of sufficient data can be handled by using DST and fuzzy analyzer in the trust computation mechanism. This computed fuzzy trust value associated with a vehicle can be updated using a reward and penalty scheme depending on the current behavior & trust worthiness. The proposed (TOEFVC) trust computation model relies on the support from cloud computing. History management can store the previous action (or initial trust value over cloud), whenever, vehicle need trust value about nearby vehicle previous trust value is attached through cloud within time limit. Hence, time factor is very important in VANET especially in safety messages. The proposed model focuses on VANET cloud. Trust computation is merging of reputation values provided by nearby vehicles.

The proposed model has three phases. Figure 1 shows the flow chart for proposed TOEFVC trust computation model. In TOEFVC trust computation has three phases. The first stage is data preprocessing with Dempster-Shafer scheme. The second stage is the fuzzy analyzer phase for trusted and untrusted vehicle. We get the end product from the second phase as an input to the fuzzy analyzer. With the help of fuzzy rules we obtained trusted and untrusted vehicles as an effect. The third phase is reward and penalty phase, if source vehicle founds that the message is right then the reward will increase gradually. Nevertheless, the punishment is severe.

4.1 Algorithm [TOEFVC Pseudo Code]

Input: Data Analysis: `[History management] Previous Level of trust reputation: Direct & Indirect) [Information as a service].`

Step1: `Create discernment frame` θ `using DST`

Step2: `Assign the basic probability assignment to each vehicle.`

Step3: `Calculate belief with the help of BPA using set related to that event (mutually exclusive).`

$$\mathrm{Bel(B)} = \Sigma_{\mathrm{V} \subseteq \mathrm{V_i}}\, \mathrm{m(V_i)}$$

`It is to calculate lower bound of confidence.`

Step4: `Calculate plausibility using set for not related to that event (mutually exhaustive)`

$$Pl\,(p) = 1 - \quad \mathrm{Bel(B)} = \Sigma_{\mathrm{V} \subseteq \mathrm{V_i}}\, \mathrm{m(V_i)}$$

`It is to calculate lower bound of confidence.`

Step5: If there is more than one vehicle for source of evidence then we used DST combiner to calculate the probability of event (trust value).

$$\mathrm{Bel(T)} = \mathrm{Bel}_1(\mathrm{V}) \oplus \mathrm{Bel}_2(\mathrm{V}_1)$$

Step6: After calculating the probability of event (trust value). We check the level of trust using fuzzy rule with the help of following formula.

Final Trust (FT) value is calculated as follows:

$$\mathrm{FT} = \frac{\Sigma(\mathrm{T}_{\mathrm{v}_1} \ldots \mathrm{T}_{\mathrm{v}_n})}{\mathrm{n}}$$

Step7: Where $\mathrm{T}_{\mathrm{v}_1}$ is the collection of trust value and set of average value of indirect trust calculated as

$$\mathrm{T}_{\mathrm{v}_1} = \left[\mathrm{T}_1 + \frac{\Sigma(\mathrm{IT}_{\mathrm{v}_1} \ldots \mathrm{IT}_{\mathrm{v}_n})}{\mathrm{n}}\right]/2$$

Step8: Here, $\mathrm{IT}_{\mathrm{v}_1}$ is the indirect trust value of $\mathrm{v}_1^{\mathrm{th}}$ vehicle. After calculating the final trust value compares this value with threshold value. If level of Trust Value is very high then vehicle comes under the category of trusted vehicle else in the category of untrusted vehicle.

Step 9: After that we apply reward for trusted vehicle and penalty for untrusted vehicle.

Step 10: Reward for trusted vehicle VCTV = VPTV + 1

Step 11: Penalty for untrusted vehicle VCTV = VPTV - 2

The above algorithm shows the whole scenario of the TOEFVC model how to execute this model represented by this algorithm. In this algorithm we take input in the form of history management using trust computation. After analyzing input we prepare the model for apply DST to fetch a probability values as form of DST result and in DST results we apply the fuzzy rules find out trusted and untrusted vehicles. In last step of this algorithm we apply reward for trusted vehicle and penalty for untrusted vehicle.

4.2 Data Preprocessing with Dempster-Shafer Scheme

Data oriented approach can be used to find out the trustworthiness of the data (for calculating the probability of event) because we know that for evaluating the trust worthiness of any vehicle first one should be sure about the event which is created or being used at time of trust calculation of any vehicle. If the events are properly created then it becomes easy to evaluate the trustworthiness of any vehicle. So here in our first phase we are going to apply Dempster-Shafer (DS) scheme [21] for calculating the trustworthiness of any event. The first step in applying the DS belief model is to define a set of possible situations which is called the frame of discernment. A frame of discernment delimits a set of possible states of a given system, exactly one of which is assumed to be true at any one time. The belief that one of the atomic states it contains is true, but that the observer is uncertain about which of them is true. Similarly to belief, an observer's disbelief must be interpreted as the total belief that a state is not true. The uncertainty state represents an observer's uncertainty regarding the truth of a given state, and can be interpreted as something that fills the void in the absence of both belief and disbelief. Dempster-Shafer theory (DST) handles uncertain and incomplete information through the definition of two dual non additive measures: plausibility and belief. These measures are derived from a density function, m, called basic probability assignment (BPA) or mass function. This probability assigns evidence to a proposition (hypothesis). To evaluate the trustworthiness of the event three probability conditions are being proposed which is based on the theory of evidence. The three used conditions taken are belief, plausibility and uncertainty. It takes direct trust values of number of vehicles with three conditions. The first condition is to have the exact information communicated by one vehicle. Second condition arises when vehicle communicate the incorrect information and in last condition vehicle is showing the sound of doubt in its transmitted message. According to Dempster theory we find all three probability conditions and put all these three conditions results in DST.

DST, also known as the theory of belief, is a simplification of the Bayesian theory of subjective probability is suitable for trust computation in the absence of infrastructure and collection of direct trust values provided by other vehicles with in the environments in VANETs. The DST owes its name to work by Dempster and Shafer [21, 22] is a numerical theory of evidence.

The belief function is defined as the

$$\mathrm{Bel(B)} = \sum\nolimits_{\mathrm{V \subseteq V_i}} \mathrm{m(V_i)} \tag{1}$$

Thus a belief function and a basic probability task express accurately the similar information equivalent to each belief function are three other commonly used quantities that express the similar information.

A function $Q : 2^{\theta \to [0,1]}$ is called a commonality function if there is a basic probability task m such that

$$\mathrm{Q(V)} = \sum\nolimits_{\mathrm{B \subseteq A}} \mathrm{m(V_1), for\,all\ V \subseteq \theta} \tag{2}$$

The doubt function is given by

$$\mathrm{Dou}(V) = \mathrm{Bel}(\neg V) \tag{3}$$

And the upper probability function is given by

$$P^{*}(A) = 1 - \mathrm{Dou}(V) \tag{4}$$

This expresses how much we can trust V if all are supporting to V.

DST combines numerous neighbor vehicles' belief on the condition that evidence from different neighbor vehicles is independent. Assuming that $bel_1(V)$ and $bel_1(V_1)$ are two belief functions over the same frame of discernment θ, the orthogonal sum of $bel_1(V)$, $bel_1(V_1)$ and $bel(V_n)$ is defined as [22]

$$bel(T) = bel_1(V) \oplus bel_2(V_1) \tag{5}$$

4.3 Data Preprocessing with Fuzzy Analyzer

After calculating the results for all three conditions involved in DST all values are taken and put in the fuzzy analyzer system. In this system we have applied some fuzzy rules on the values obtained by first phase. These fuzzy rules have some threshold value which decides the level of trusted or untrusted vehicle. Where the Function for deciding threshold values of fuzzy rules depends upon past information of vehicle itself. The range for the trusted or untrusted vehicle using fuzzy rules is 0–1. With the help of our system the accuracy rate of correct answer is 96.5%. Accuracy rate is being calculated by testing the whole system on various scenarios. In the proposed scheme [3, 22] trust decision is based on fuzzy logic.

After calculating the probability of event (or trust values), we check the level of trust using fuzzy rule with the help of following formula.

Final Trust (FT) value is calculated as follows:

$$FT = \frac{\sum(T_{V_1}\ldots\ldots T_{V_n})}{n} \tag{6}$$

Where T_{V_1} is the collection of trust value and set of average value of indirect trust (IT) calculated as

$$T_{V_1} = \left[T_1 + \frac{\sum(IT_{V_1}\ldots\ldots IT_{V_n})}{n}\right]/2 \tag{7}$$

Here, IT_{V_1} is the indirect trust value of v_1^{th} vehicle. So if the computed trust is equal or greater than the threshold value then that vehicle comes under the category of trusted vehicle otherwise not. After that on the basis of trustworthiness of vehicle obtained by fuzzy analyzer scheme the vehicle can be capable of doing each and every task give to it.

Trust levels based on fuzzy logic are divided as trustworthy, very trustworthy, medium trustworthy, untrustworthy and very untrustworthy [5] (Table 1).

Table 1. Fuzzy discrimination

S. no.	Fuzzy levels	Trust values	Semantics
1.	Very high	0.8 to 1	Trustworthy
2.	High	0.6 to 0.8	Trustworthy
3.	Medium	0.5 to 0.6	Trustworthy
4.	Low	0.2 to 0.5	Untrustworthy
5.	Very low	0 to 0.2	Untrustworthy

In this phase, fuzzy rules are applied on the values obtained from first phase. Trust is considered a measurable continuous subjective quantity. Trust quantity has a minimum value (that means having no trust) and a maximum value (that means completely having trust) where partial trust comes in-between. We have taken 0 as the lower and 1 as the higher bounds of trust results the range of [0, 1] for trust.

Fuzzy modeling of trust where trust values are represented as fuzzy sets with definite membership functions in the trust range. However, trust is considered a measurable continuous subjective quantity. The fuzzy membership functions for the linguistic terms such as low, medium, medium low, medium high, and high can be defined as depicted in Fig. 2a. Hence, the linguistic terms can also be used as trust values. Figure 2b the fuzzy set approach to the set of trusted vehicle provides a representation of the trustability of the vehicle.

Fuzzy small and large values calculated with the help of fuzzification algorithms of the member of the fuzzy set. In this algorithm we find the mid are subjectively given by the vehicle's message. However, algorithm will not address negative trust values. So the trust values must in positive values before fuzzification [23]. So fuzzification algorithm for small or large value defined as:

$$\mu(x) = \frac{1}{1 + \left(\frac{x}{V2}\right)^{V1}} \tag{8}$$

$$\mu(x) = \frac{1}{1 + \left(\frac{x}{V2}\right)^{-V1}} \tag{9}$$

Where V_1 is the spread of the transition from a membership value of 1 to 0 and V_2 is the midpoint where the membership value is 0.5. Here $\mu(x)$ is the membership function.

In the proposed scheme [24, 25] trust decision is based on fuzzy logic. If the computed trust is equal or greater than the threshold value then that vehicle comes under the category of trusted vehicle otherwise not. Trust levels based on fuzzy logic are divided as trustworthy, very trustworthy, medium trustworthy, untrustworthy and very untrustworthy [5]. Trust levels are considered based on exchange messages among

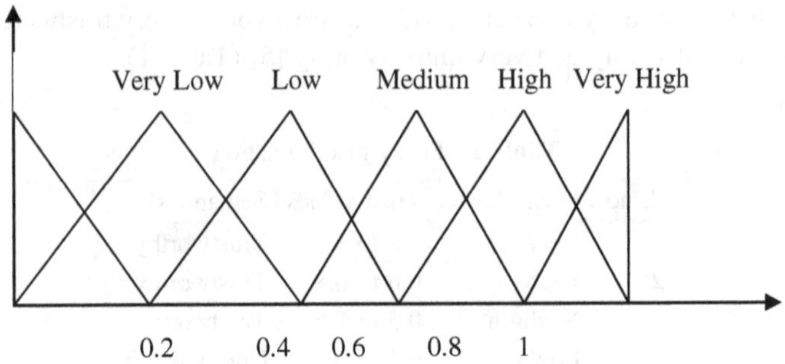

(a) Fuzzy membership functions of trust linguistic terms

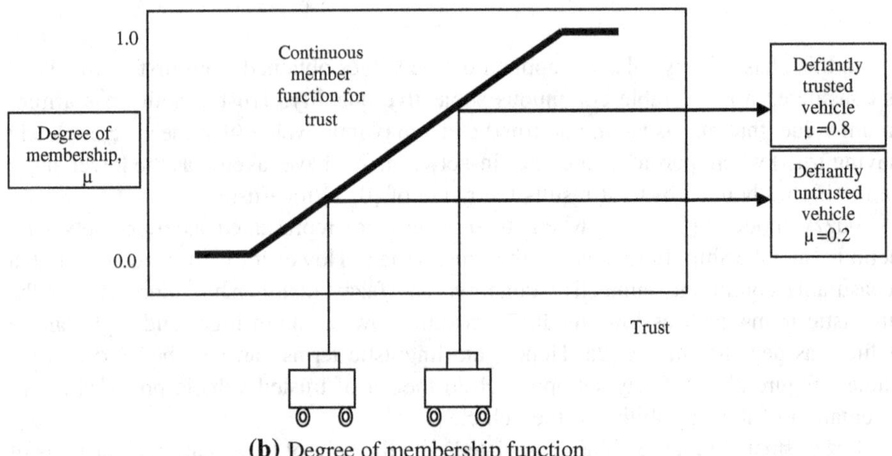

(b) Degree of membership function

Fig. 2. (a) Fuzzy membership functions of trust linguistic terms (b) Degree of membership function

Table 2. Trust level based on message classification in VANET

Message category	Example	(Min/Max) Trust value requirement	For message channel assignment priority
Safety critical messages	Crash-pending notification, hard-brake (collision warning, EEBL, LANE changing, anti-lock, etc.) and control loss	0.8–1.0	Very High
Safety related messages	Emergency vehicle approaching, probable-situation (e.g., rapidly deteriorating dangerous conditions), signal phase and timing (SPAT)	0.6–0.8	High
Non-safety application	Electronic toll tax collection	0.6	Moderate
Non-safety messages	Location-based services: car parking information, off-board navigation reroute instructions, location finding	0.5–0.6	Low
Non-safety messages	Infotainment services: digital map download, songs and games downloading services	0.5	Very Low

vehicles in VANET. Messages are categorized as safety message (Crash-pending notification, hard-brake, collision warning, and Emergency vehicle approaching and probable-situation) and non safety messages (Location-based services, car parking information, Electronic toll tax collection, digital map download, songs and games) so we use in this work computation of trust using fuzzy rules are extended in five level based on messages. Trust level based on messages exchange among vehicles in VANET depicted by Table 2.

4.4 Reward and Penalty Scheme

In this phase, we have applied an algorithm to give rewards to trusted vehicle and penalty for untrusted vehicle. This proposed mechanism works without any help of certificate authority (CA) or any road side infrastructure. If any vehicle misbehaves form the starting point then it cannot be able to do any kind of damage in reality because the vehicle joins with the system and at fuzzy analyzer system check the trustworthiness of vehicle at every moment. [26]. If it tries to gain some credit and then misbehave the rewards and penalties system will recognize a misbehaving strike and punish it. So for each misbehaving vehicle we are applying penalty of −2 and reward of +1 for trustworthy vehicle is shown below by algorithm

4.4.1 Algorithm - Pseudo Code for Rewards and Penalty

```
Step1 : (Rewards): Do {                    // If found trusted vehicle then rewards
Step2: If (not misbehaving) then           // i.e. vehicle is not malicious Table2
Step 3: Type [VT |VI | VPTV]               // Type shows, vehicle type, vehicle identity and
Step 4: VCTV = VPTV + 1 else               // vehicle previous trust value, so value added + 1
Step5: (penalty): Type [VT|VI|VPTV]// If found untrusted vehicle then penalty apply
Step 6: VCTV = VPTV - 2    } while (true);
                                           // vehicle previous trust value, so value subtract by 2
Step 8: Update trust table or go to step 2;
                                           // finally, trust table updated
```

Algorithm shows the reward and penalty computed by fuzzy analyzer. Here, vehicle type (VT), vehicle identity (VI), vehicle previous trust value (VPTV) and vehicle current trust value (VCTV) respectively. If the source vehicle found that the message is transmitted by trusted vehicle and so the trust value will increase by 1, if the source vehicle finds that false message by misbehaving vehicle then the trust value will fall by 2. In the proposed trust computation scheme trust value increase gradually while trust value of message decrease rapidly. In next section, simulation and results is presented.

5 Simulation and Results

In this section, we introduce the detailed TOEFVC model simulation setup and results, The efficacy of proposed model in VANET with cloud analyst [27], CloudSim [28], NetBean [29] and VANETsim [30]. This is open source java based visual modeler tool for analyzing cloud computing application. It is to design to simulate scalable cloud application with the purpose of simulation and behavior of such application example proposed work calculation and behavior to analyze how to reduce death rate or find trustworthy nodes. CloudAnalyst helps developers with insights in how to distribute applications among cloud infrastructures and value added services such as optimization of applications performance and incoming providers with the use of service brokers [31], depicted by Fig. 3.

5.1 Simulation Setup

To verify the efficacy of the proposed trust computation model, its operation is evaluated under actual traffic conditions on the road with cloud analyst and CloudSim. We create a scenario with help of cloud analyst after the analysis of benchmark dataset and then with these scenario active vehicles, average speed (Km/h), average travel distance (m), average travel time(s), vehicles, average known vehicles, created unique messages, failed forward messages. After that we have designed a TOEFVC model for computing the trusted or untrusted vehicles by taking the values obtained from simulation. We apply DST on the values obtained while creating a scenario and apply fuzzy rules on those values which we obtained as an outcome of DST to computing the trust value of messages of vehicles.

Internet configuration and running state for configuration describes the transmission regions and units in milliseconds. Our configuration using data values, data center configuration, simulation duration, number of vehicles and service broker policy. Cloud simulator [29] describes data center, virtual machine, image size, memory and bandwidth all components are required for effective simulation in TOEFVC model for trust computation.

Simulator create a cloud broker with the help of cloud broker we create the user base. User base is based on the clusters of vehicles. After creation of user base we assign the virtual machine to each user base. Cloud analyst generates all the results such that message sent, message received and processing time etc. The proposed simulation using cloud stores and retrieves data in a shorter time than the traditional VANET environment. Here we used CloudAnalyst for fetching and retrieving data. It is very fast in terms of data transmission, cost and time.

5.2 Results and Discussion

From Fig. 3(a, b, and c) we observe user base hourly response times for a different number of vehicles. Here we select the group of low, medium and dense vehicle environment and simulate the result.

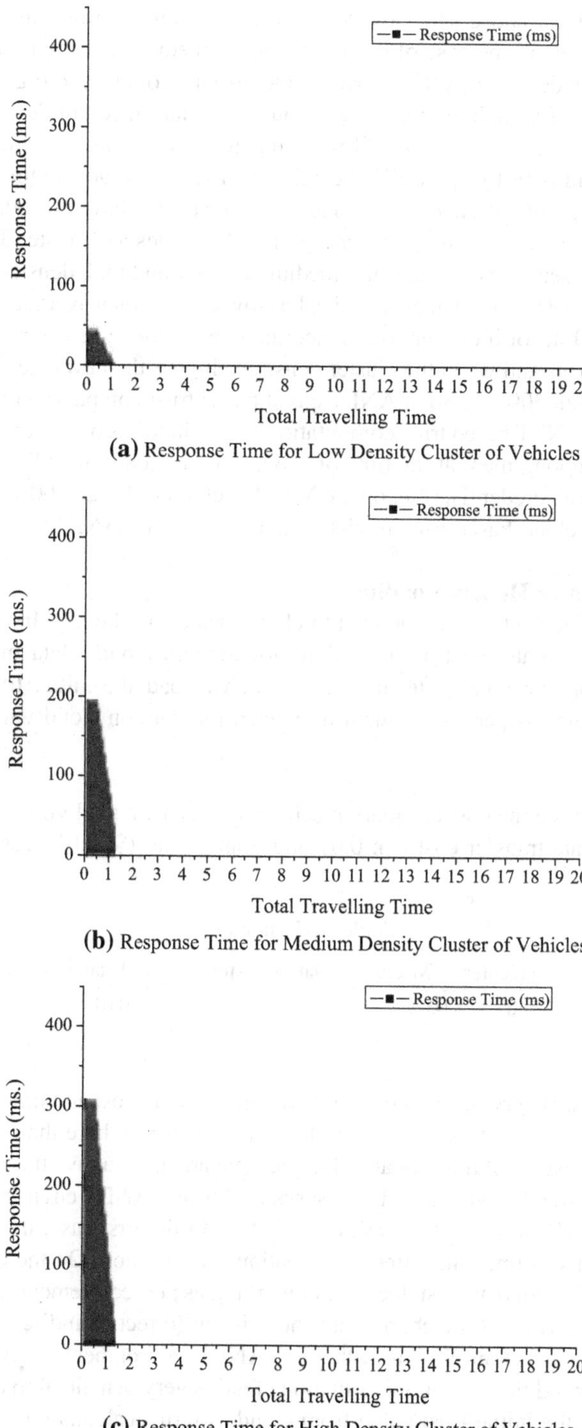

Fig. 3. (a) Response time for low density cluster of vehicles (b) Response time for medium density cluster of vehicles (c) Response time for high density cluster of vehicles

However, Fig. 3a and c show the low density of vehicle group in terms of time response of very fast in the case of safety messages. medium density of vehicle group comparison to low density of vehicle group time response of fast and Fig. 3b shows the high dense group of vehicle group comparison to low and medium density of vehicle group time response of little fast. But if we compare cloud environment into non cloud environment cloud is fast response in the case of safety messages. Figure 3 a, b and c show the response time of clusters 1, 2 and 3. However, we have considered cluster 1 has group of 50 vehicles, cluster 2 has group of 200 vehicles and cluster 3 has group of 300 vehicles represented as low density, medium density and high density respectively.

It is also observed that when cluster size has low density then average response time is nearly about 60 ms or if cluster size is medium density then average response time is nearly about 200 ms and when the cluster density is higher than average, then response time in nearly about 300 ms. So, VANET cloud based trust computation model is good with respect to VANET based trust computation model in terms of response time during communication among them at the time of driving on the road. In addition, all clusters may upload security credentials etc. on the VANET cloud less than 600 ms. This shows that the VANET cloud based trust model is more viable in VANET.

5.2.1 Data Center Hourly Loading

The user base response time is uploaded to cloud data center hourly. In this section all groups regularly update the trust related information in cloud. Data management is important part of any type of technology. Here we load the initial trust value and updated trust values as per requirement user fetch the data on cloud within time.

5.2.2 Cost

In this section we see the cost calculation using cloud analyst total virtual machine cost ($): 0.01, total data transfer cost ($): 0.01 and grand total: ($) 0.01 (Table 3).

Table 3. Data cost

Datacenter	VM cost $	Data transfer cost $	Total $
Cloud1	0.01	0.01	0.01

Cloud Analyst is very fast in terms of data transmission, cost and time. With help of cloud we store the initial trust values on cloud. It is observed here that cloud provides data as a service instead of traditional techniques that are used to give this service. Cloud is very beneficial for VANET cloud as a service. But in VANET environment, we talk about two cases offline and online (V2I and V2V). In the first case, the cloud is very beneficial in terms of time and safety information transmission. On the other hand the second case (V2V) cloud analyst doesn't not working as per requirement. So we apply in case 1 and case 2 (from flow chart) both the scheme (direct & indirect) models with cloud for trust management. We know that in VANET environment safety messages are transferred in limited time. So in this situation cloud is very beneficial to calculate trust. Here we fetched the initial trust value from cloud. So trust calculation is fast.

5.3 Results for Data Preprocessing with Dempster-Shafer Scheme

In this simulation, [22] we have considered basic probability task (BPT) value of three vehicles V_1, V_3 and V_8. It is also observed that these values are less than average value of BPT. To evaluate the set values from power set of three vehicles using step 2 and step 3 from DST phase 1, Algorithm 4.1 (Fig. 4).

$$V = \{V_1, V_3, V_8\}$$

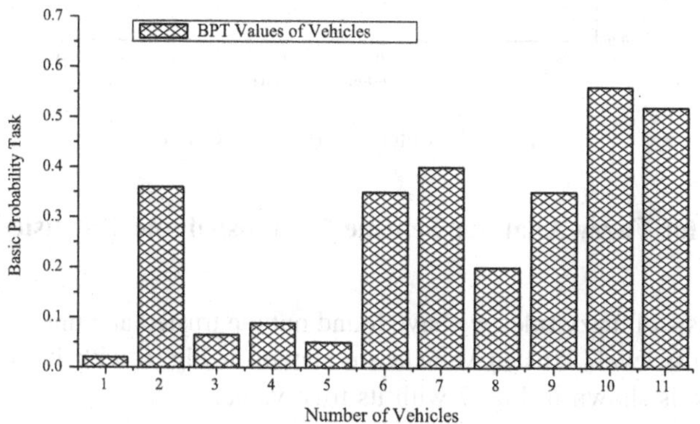

Fig. 4. The basic probability task for each vehicle(s)

We have assumed the values of all possible sub sets.

$$V_1 = \{0.02\}, \{V_1, V_3\} = 0.05 \text{ and } \{V_1, V_3, V_8\} = 0.20$$

The values of beliefs evaluated by proposed models from Eq. 1 are Bel $(\{V_1\}) = 0.02$, Bel $(\{V_1, V_3\}) = 0.07$, Bel $(\{V_1, V_8\}) = 0.002$ and Bel $(\{V_1, V_3, V_8\}) = 0.27$.

After calculating belief of vehicle, we have calculated the value of commonality function Q(V) from DST.

$$Q(\{V_1\}) = 0.27, Q(\{V_1, V_3\}) = 0.25, Q(\{V_1, V_8\}) = 0 \text{ and } Q(\{V_1, V_3, V_8\}) = 0.20$$

We have calculated doubt value from DST $\text{Dou}(V_1) = 0$.

From DST, we have evaluated final value of belief Bel $= 0.68$. Hence, belief value is close to 1, so vehicle can take trust upon received location based information from other vehicles within its vicinity. In Fig. 7 we have seen the outcome of the basic probability task of belief function. After getting these values we put all these values in DST for further processing and the outcome of DST is shown in Fig. 5.

It shows the probability of event in this figure we see that number of vehicle and probability values lie amid 0 to 1. If values are closer to 1 then we can say that occurrence of event is high otherwise low using fuzzy rules analyzer phase 2.

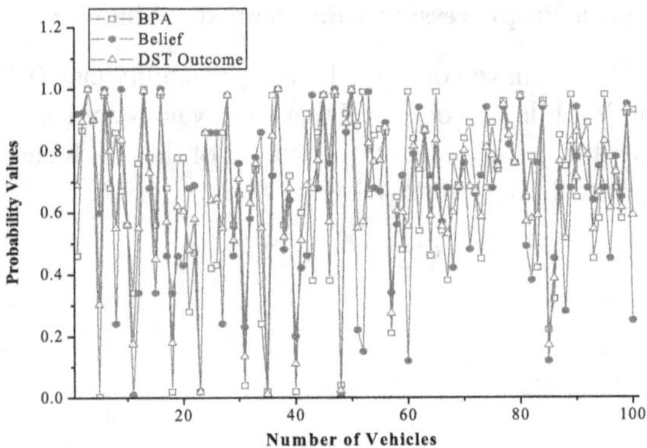

Fig. 5. The outcomes of BPA & belief

5.4 Results for Fuzzy Analyzer Scheme for Trusted and Untrusted Vehicle

Here on the basis of fuzzy rules we have found out the trusted and untrusted vehicle on the basis of trust level. Trusted vehicle are shown in Fig. 6 with its trust value and untrusted value is shown in Fig. 7 with its trust value.

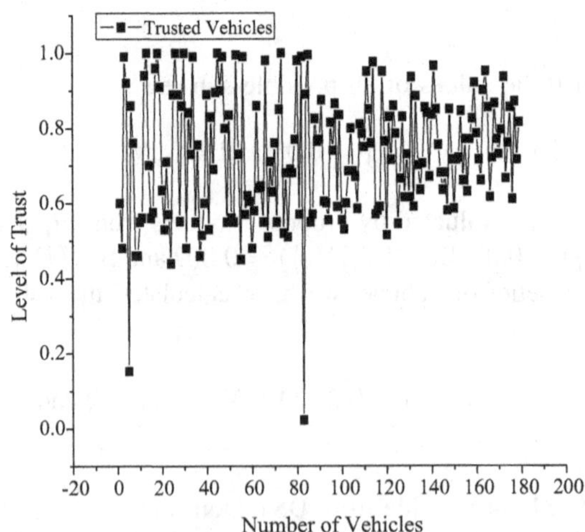

Fig. 6. Trusted vehicle as an outcome of fuzzy analyzer

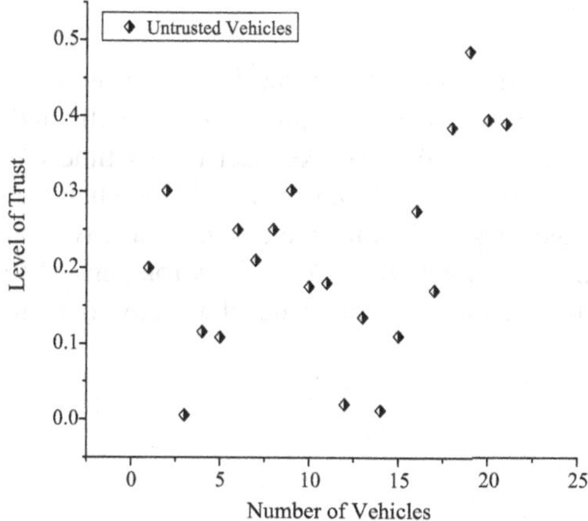

Fig. 7. The untrusted vehicles as an outcome of fuzzy analyzer

Figures 6 and 7 show the outcome of phase 2. With the help of fuzzy analyzer we found out trusted and untrusted vehicle as an outcome using fuzzy rules. It shows the level of trust. We see that number of vehicle and its trust values lie amid 0 to 1. If values closer to 1 then we can say that vehicle is trusted otherwise untrusted. After find out level of trust we apply reward and penalty scheme phase 3.

5.5 Results for Reward and Penalty Scheme

Here we have seen an outcome of the third phase of an algorithm. The graph shows the updated value of the trust level of vehicle after applying reward of +1 and penalty of −2. As it is known to all that creation of trust goes in slow motion but trust destruction takes very less time (Fig. 8).

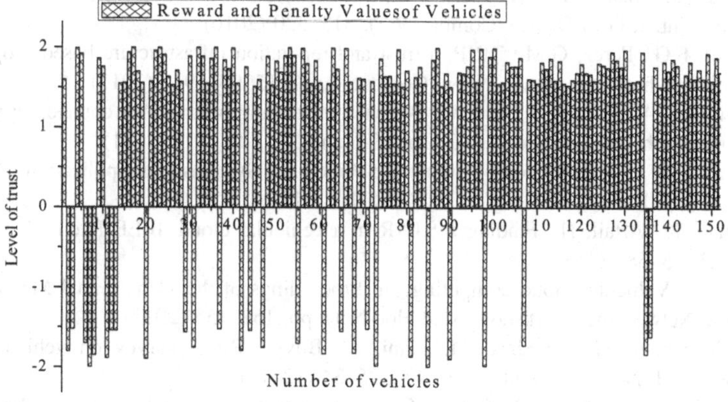

Fig. 8. The level of trust after applying reward & penalty

6 Conclusion

The paper addressed the trust computation mechanism using cloud computation. The proposed TOEFVC model was able to compute trust correctly and quickly in VANET environment. The vehicle was able to take decision in time constrained ephemeral environment. The reward penalty mechanism was able to change trust value as needed in VANET for message dissemination in the vehicular environment. The proposed TOEFVC mechanism fits well with the VANET environment when smart vehicle have hard delay constraints and high mobility and cloud environment may only provide suitable candidature.

References

1. Li, L., Song, J., Wang, F.-Y., Niehsen, W., Zheng, N.-N.: IVS 05: new developments and research trends for intelligent vehicles. IEEE Intell. Syst. **20**, 10–14 (2005)
2. Boban, M., Vinhoza, T.T.V., Ferreira, M., Barros, J., Tonguz, O.K.: Impact of vehicles as obstacles in vehicular ad hoc networks. IEEE J. Sel. Areas Commun. **29**(1), 15–28 (2011)
3. Khabazian, M., Aissa, S., Ali, M.M.: Performance modeling of safety messages broadcast in vehicular ad hoc networks. IEEE Trans. Intell. Trans. Syst. **14**(1), 360–387 (2013)
4. Soni, S., Sharma, K., Chaurasia, B.K.: Trust based scheme for location finding in VANETs. In: Lakshminarayanan, V., Bhattacharya, I. (eds.) Advances in Optical Science and Engineering. SPP, vol. 166, pp. 425–432. Springer, New Delhi (2015). https://doi.org/10.1007/978-81-322-2367-2_53
5. Wei, Y.C., Chen, Y.M.: Adaptive decision making for improving trust establishment in VANET. In: 16th Asia-Pacific Network Operations and Management Symposium (APNOMS), pp. 1–4 (2014)
6. Cho, J.-H., Swami, A.: Towards trust-based cognitive networks: a survey of trust management for mobile ad hoc networks. In: Proceedings of the 14th International Command and Control Research and Technology Symposium, pp. 1–16 (2009)
7. Gerlach, M.: Trust for vehicular applications. In: Eighth International Symposium on Autonomous Decentralized Systems, pp. 295–304. IEEE Computer Society (2007)
8. Sharma, K., Chaurasia, B.K., Verma, S., Tomar, G.S.: Token based trust computation in VANET. Int. J. Grid Distrib. Comput. **9**(5), 313–320 (2016)
9. Marmol, F.G., Perez, G.M.: TRIP, a trust and reputation infrastructure based proposal for vehicular ad hoc networks. J. Netw. Comput. Appl. **35**, 934–941 (2012)
10. Benslimane, A., Barghi, S., Assi, C.: An efficient routing protocol for connecting vehicular networks to the internet. Pervasive Mob. Comput. **7**(1), 98–113 (2011)
11. Gillam, L., et al. (eds.): Cloud Computing: Principles, Systems and Applications. Springer, London (2010). https://doi.org/10.1007/978-1-84996-241-4
12. Mershad, K., Artail, H.: Finding a STAR in a vehicular cloud. IEEE Intell. Transp. Syst. Mag. **23**, 55–68 (2013)
13. Gerla, M.: Vehicular cloud computing. In: Proceedings of the 11th Annual Mediterranean Ad-Hoc Networking Workshop (Med-Hoc-Net), pp. 152–155 (2012)
14. Whaiduzzaman, M., Sookhak, M., Gani, A., Buyya, R.: A survey on vehicular cloud computing. J. Netw. Comput. Appl. **40**, 325–344 (2014)
15. Mousannif, H., Khalil, I., Olariuc, S.: Cooperation as a service in VANET: implementation and simulation results. Mob. Inf. Syst. **8**(2), 153–172 (2012)

16. Chen, B., Curtmola, R., Ateniese, G., Burns, R.: Remote data checking for network coding-based distributed storage systems. In: Proceedings of the 2010 ACM Workshop on Cloud Computing Security Workshop, pp. 31–42 (2010)
17. Chaurasia, B.K., Verma, S.: Secure pay while on move toll collection through VANET. Int. Comput. Stand. Interfaces **36**(2), 403–411 (2014)
18. Song, S., Hwang, K., Zhou, R., Kwok, Y.-K.: Trusted P2P transactions with fuzzy reputation aggregation. IEEE Internet Comput. **9**(6), 24–34 (2005)
19. Griffiths, N., Chao, K.-M., Younas, M.: Fuzzy trust for peer-to-peer systems. In: 26th IEEE International Conference on Distributed Computing Systems Workshops (ICDCS 2006), pp. 73–79 (2006)
20. Josang, A.: A logic for uncertain probabilities. Int. J. Uncertain. Fuzziness Knowl.-Based Syst. **9**, 279–311 (2001)
21. Dempster, A.P.: A generalization of Bayesian inference. J. R. Stat. Soc. Ser. B **30**, 205–247 (1968)
22. Shafer, G.: A Mathematical Theory of Evidence. Princeton University Press, Princeton (1976)
23. Chen, C., Zhang, J., Cohen, R., Ho, P.-H.: A trust-based message propagation and evaluation framework in VANETs. In: Proceedings of the International Conference on Information Technology Convergence and Services, pp. 1–8 (2010)
24. Hao, Y., Cheng, Yu.: A distributed key management framework with cooperative message authentication in VANETs. IEEE J. Sel. Areas Commun. **29**(3), 616–629 (2011)
25. Rawatz, D.B., Bistax, B.B., Yan, G., Weigley, M.C.: Securing vehicular ad-hoc networks against malicious drivers: a probabilistic approach. In: International Conference on Complex, Intelligent and Software Intensive Systems, pp. 146–151 (2011)
26. Rivas, D.A., Zapata, M.G.: Chains of trust in vehicular networks: a secure points of interest dissemination strategy. J. Netw. Comput. Appl. **10**(6), 1115–1133 (2012)
27. Olariu, S., Khalil, I., Abuelela, M.: Taking VANET to the clouds. Int. J. Pervasive Comput. Commun. **7**(1), 7–21 (2011)
28. Hasan, R.: Cloud security. www.cs.jhu.edu/ragib
29. opensourceforu.com/…/cloudsim/
30. www.vanet-sim
31. Denko, M.K., Sun, T., Woungang, I.: Trust management in ubiquitous computing: a Bayesian approach. Comput. Commun. **34**(3), 398–406 (2011)

Received Power Analysis of Cooperative WSN Deployed in Adjustible Antenna Height Environment

Sindhu Hak Gupta(✉) and Niveditha Devarajan

Amity School of Engineering and Technology, Noida, India
shak@amity.edu, nivedith.dev@gmail.com

Abstract. The work in this paper aims at showing the effectiveness of cooperative communication considering a Wireless Sensor Network (WSN) scenario deployed in a snowy environment. Sensor nodes with adjustable antenna heights are assumed so that even when the ground is covered with snow, these antenna nodes are able to transmit data. Received power of the deployed cooperative network with adjustable antenna heights has been evaluated and further using it as a crucial parameter a comparative analysis of cooperative and non-cooperative scenario is carried out. An Amplify and Forward cooperative protocol has been considered for modeling the network and deriving the expressions for energy and power. Further, various path loss models have been considered to compute the pathloss and evaluate the performance of network. Simulation and analytical results reveal that increasing the antenna height reduces the received power and cooperative scenario outperforms the non-cooperative scenario.

Keywords: Wireless Sensor Network · Cooperative communication · Amplify and forward · Path loss · Received power

1 Introduction

Wireless Sensor Network (WSN) consists of miniature sized, battery dependent, low power, low storage computer like devices called nodes, working mutually to form the network. These sensor nodes on being deployed work as a separate, independent network for sharing crucial information. The nodes are capable of gathering, processing and data computation [1]. The parameters detected by these sensors include humidity, motions, temperatures etc. [2].

The structure of the nodes enables them to be placed in severe, difficult environments where placement of other communication structures may not be possible and hence may cause a failure in sensing and communication [3]. WSNs have therefore been utilized in many crucial applications including environmental monitroing and rescue operations. It is being utilized industrially and commercially due to its expanding advancements in technology. Overwhelming research work has been done where reliability and functioning of a WSN has been analyzed [4]. Reliability and Energy Efficiency are two critical parameters required for successful operation of a

M. L. Gavrilova and C. J. K. Tan (Eds.): Trans. on Comput. Sci. XXXIV, LNCS 11820, pp. 96–108, 2019.
https://doi.org/10.1007/978-3-662-59958-7_6

WSN. Reliability makes sure that accurate data reaches the destination and enhancing the energy efficiency increases the lifetime of a WSN. Since, the nodes are memory limited, it becomes necessary that correct data reaches the destination node. Issues like multipath channel fading increases path loss, reducing the reliability of transfer of information. Along with the multipath fading, the climatic conditions may affect the devices. Power of sensor node is consumed during the communication process that reduces the lifetime of the network. Therefore, direct communication between nodes placed in extreme climatic conditions and environment may not be feasible. Cooperative communication has proved to be a beneficial solution to the multipath fading in wireless communication. The coordinated effort of sensors increases reliability of link, throughput, diversity order and reduces consumption of energy, increasing longevity of WSN [5, 6]. Cooperative communication involves transmitting information from source node to the destination node through relay nodes. The relay node aggregates the information and furthers the information to destination node, implementing multihop communication. This makes multiple independent paths for the transfer of data [7, 8]. In the light of tasks performed by relay, the cooperative signaling relaying protocols can be mainly broken into Amplify and Forward protocol, Decode and Forward protocol and Coded Cooperative communication. Here, Amplify and Forward techniques are considered for simple and uncomplicated implementation in WBAN [9–12]. Communication channel is stochastic and arbitrary in nature, there are several problems like interference, scattering due to the nature of the channel might attenuate the data transfer through it. Therefore, pathloss has to be predicted mathematically to evaluate performance of the network and improve communication. Path loss models aid in predicting path loss received by transmitted data to the destination node. An accurate prediction method aids in optimizing the power needed for transmission, area covered and reduces interference [13–16]. Received Energy analysis of a WSN with constant antenna height nodes has been explored for cooperative and non-cooperative scenario [19]. In the current work, various path loss models such as Free Space path loss model, Empirical Log-Distance model, Two-Ray model and Ray- Tracing model [4] are being considered for evaluating and analyzing the effectiveness of cooperative communication of WSN in an adjustable height scenario.

The deployment and working of a WSN in a snowy environment is different from the conventional environments like a normal farmland. As, when the snow falls, it covers the ground making the necessary sensor reading and further transmission a challenge. *Cheffena et al.* [4] deployed a sensor network in a snowy environment and studied the effects of path loss on the network. For doing so, he studied the path loss effects varying the antenna heights assuming that with time, the magnitude of snow is increasing with respect to the ground. Since cooperative communication has proved beneficial in multipath scenario, in the current work, cooperative communication is implemented in a WSN scenario with adjustable antenna heights. To investigate the effect of cooperative communication and variation in antenna heights, mathematical modelling has been done to derive expressions for received energy and power at the destination node for the cooperative case. The received power has been observed for the critical comparitive analysis of cooperative and non-cooperative scenario. This observation is for three antenna heights with respect to the ground.

For the current work, the sensors have been assumed to be deployed in a snowy field. To practically analyze the effect of cooperative communication, power received at the destination node for cooperative as well as non-cooperative case is modelled and simulated with the aid of various path loss models including the measured values taken by Cheffena. Amplify and forward protocol has been implemented to mathematically derive expressions for received energy and power for the cooperative case. The evaluation has been done for varying antenna heights. The heights of antenna of sensor nodes have been varied to obtain proper measurements of path loss as the signals from short heighted nodes might get buried under the snow, not giving appropriate values. The highlight of the work is that the results will present a comparative demarcation in the received power at the destination node for cooperative and non-cooperative WSN scenario with adjustable antenna heights. Amplify and forward cooperative relaying strategy and pathloss models have been to derive expressions for cooperative energy and power evaluation.

The rest of the work in this paper is divided into four sections: Sect. 2 explains the WSN model, Sect. 3 elaborates the results obtained and the paper is concluded in Sect. 4.

2 System Model

The WSN setup has been assumed to be deployed in a snowy field. Practically the network will consist of multiple nodes being deployed in the field for but for our simplicity of analysis, three nodes being source (S), destination (D) and the relay node (R) with adjustable antenna heights are considered. These nodes are kept at three heights (0.25 m, 1 m and 1.5 m) from the ground to find out the path loss values. Destination node's position is changed from 5 m to distance of 30 m with a changing

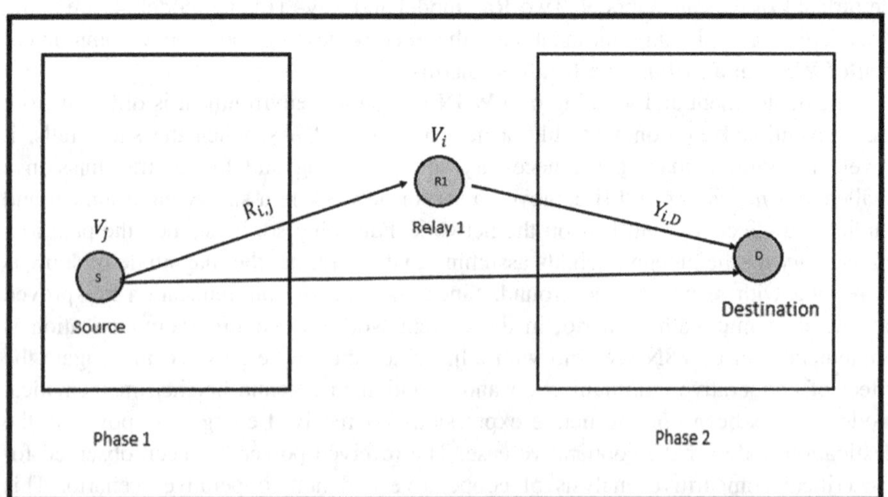

Fig. 1. Model for WSN

step size of 5 m and values of path loss and further, energy and power were calculated at these distances. The Amplify and Forward protocol has been considered. The relay node aids in amplification of data from source and furthers the information to the destination node as shown in Fig. 1. It is assumed that no power losses occur during communication.

Given in Table 1 are the important notations and the terminologies required in the system model for evaluating the performance.

Table 1. Notations and terminology

$R_{i,j}$	Signal received at relay node
e_{s1}	Symbol Power in first stage (dBm)
μ	Amplification factor at the relay
e_{s2}	Symbol Power in second stage (dBm)
N_ρ	Noise Power spectral density (dBm/Hz)
r	Channel capacity (bits/second)
PL	Path loss (dB)
d	Distance between source/relay node and destination node

Stages of Communication

Stage 1

Considering the first stage, the source node (S) after receiving and sensing the information shares or transmits this information with the neighboring node (R) that acts as the relay. This scenario has been analyzed in [17]. The node receives this data and store it in its memory. The received $R_{i,j}$ signal is given in (1) and has been represented in Fig. 1.

$$R_{i,j} = \sqrt{e_{s1}} v_j \beta_{i,j} + n_{i,j} \tag{1}$$

where e_{s1} is transmission symbol power in stage 1, v_j is a BPSK-modulated symbol sent from node V_j, $\beta_{i,j}$ is the amplification factor and $n_{i,j}$ is additive white Gaussian noise at node i from node j, having variance N_δ.

Stage 2

In second stage or in the relay transmit stage, the relay furthers the amplified signal or data to destination node. The signal is amplified by an amplification factor $\mu_{i,j}$. The information or message required to be set from a node is small, the sensors being Energy constrained don't send the information directly to the destination nodes as it will be Energy inefficient. Therefore, aggregation of information is used at the relay to save Energy as well as to reduce the information packet overhead. The data aggregation involves collection of data at an intermediate stage (the relay) that has been sent by the source node or nodes involved in cooperation.

The relay node V_i combines received information which has been amplified by the factor $\mu_{i,j}$, and its own information v_i. This information combined forms the cooperative data packet x_i given in (2).

$$x_i = \begin{cases} \mu_{i,j}\hat{u}_{i,j} & j \neq i \\ v_i & j = i \end{cases} \quad (2)$$

After the combination, the relay node sends the cooperative data packet to destination node.

The received signal $Y_{i,D}$ at the destination node is mentioned in (3) and has been shown in Fig. 1.

$$Y_{i,D} = \sqrt{e_{s2}} x_i \alpha_{i,D} + n_{i,D} \quad (3)$$

where e_{s2} is the symbol power transmitted in stage 2, $\alpha_{i,D}$ is log normal fading channel constant, $n_{i,D}$ is additive white Gaussian noise at destination D from node i, having power spectral density N_ρ.

Equation (3) has been expanded further and is given by (4).

$$Y_{i,D} = \frac{\sqrt{e_{s1} e_{s2}}}{\sqrt{e_{s1} \left|\beta_{i,j}\right|^2 + N_o}} \alpha_{i,D} \beta_{i,j} v_i + n'_{i,D} \quad (4)$$

The expression for $n'_{i,D}$, taken from (4) is expressed in (5).

$$n'_{i,D} = \frac{\sqrt{e_{s2}}}{\sqrt{e_{s1} \left|\beta_{i,j}\right|^2 + N_o}} \alpha_{i,D} n_{i,j} + n_{i,D} \quad (5)$$

The Shannon Hartley theorem expresses the highest rate at which the information is sent over a communication channel with a particular or fixed bandwidth B and in the presence of noise. The mathematical expression is given in Eq. (6).

$$r = B \log_2 \left(1 + \frac{S}{N} \right) \quad (6)$$

Here r denotes channel capacity in bits per second, S denotes average signal power, N denotes average Power of noise and interference, B stands for bandwidth. It is also known from (7) that

$$N = BN_o \quad (7)$$

where, N_o indicates noise power spectral density, B indicates bandwidth.

Substituting above relation in (6) and replacing S by e_{s1}, (8) is obtained as below

$$r = B \log_2 \left(1 + \frac{e_{s1}}{BN_o} \right) \quad (8)$$

From (8) the value of N_o can be furthered as:

$$N_o = \frac{e_{s1}}{(2^{\frac{r}{B}} - 1)B} \tag{9}$$

Substituting the value of $n'_{i,D}$ from (5) in (4) to get (10)

$$Y_{i,D} - n_{i,D} = \frac{\sqrt{e_{s2}}}{\sqrt{e_{s1}\left|\beta_{i,j}\right|^2 + N_o}} \alpha_{i,D}\left(\beta_{i,j} v_i \sqrt{e_{s1}} + n_{i,j}\right) \tag{10}$$

Comparing (10) with (3), to obtain (11) given below

$$x_i = \frac{\sqrt{e_{s1}}\beta_{i,j} v_i + n_{i,j}}{\sqrt{e_{s1}\left|\beta_{i,j}\right|^2 + N_o}} \tag{11}$$

Substituting the value of x_i from (11) in (3), (12) is obtained

$$Y_{i,D} - n_{i,D} = \sqrt{e_{s2}}\alpha_{i,D} \frac{\sqrt{e_{s1}}\beta_{i,j} v_i + n_{i,j}}{\sqrt{e_{s1}\left|\beta_{i,j}\right|^2 + N_o}} \tag{12}$$

From (12), finding the expression for e_{s2} by substituting the value of N_o from (9), to obtain (13).

$$e_{s2} = \frac{(Y_{i,D} - n_{i,D})^2 \left[e_{s1}\left|\beta_{i,j}\right|^2 + \frac{e_{s1}}{(2^{\frac{r}{B}}-1)B}\right]}{\alpha_{i,D}^2 \left[\sqrt{e_{s1}}\beta_{i,j} v_i + n_{i,j}\right]^2} \tag{13}$$

Equation (13) relates the transmitted symbol power e_{s2} with terms like channel rate r, bandwidth B, amplification factor $\alpha_{i,D}$ at the relay.

Path loss models play a very crucial role in the performance evaluation of a WSN. These models assist in predicting the attenuation of the signal between the sender and the receiver node. Attenuation is a function of propagation distance and other various other factors depending upon the type of model used. Various path loss models have been considered for the evaluation and analysis of received power at the destination node. The path loss models that are considered as given as follows:-

Free Space Path Loss Model: This evaluates the loss in strength and signal power of an EM wave coming about from LOS(Line of Sight) path through free space with no deterrents nearby to cause diffraction or reflection. The equation in dB is computed in [4] and shown by (14).

$$PL_{free} = 20\log_{10}\left(\frac{4\pi fd}{c}\right) \tag{14}$$

where f indicates frequency in MHz and c denotes speed of light. The expression in (14) is written as shown in (15).

$$PL_{free} = -27.56 + 20log_{10}(f) + 20log_{10}(d) \tag{15}$$

Equation (15) can further be written in terms of Energy (E), utilizing Planck's equation written as ($E_J = h_{J/Hz} f_{Hz}$). With the help of Planck's equation, (15) is expressed as (16).

$$PL = 636.01 + 20log_{10}(E) + 20log_{10}(d) \tag{16}$$

Two Ray Path Loss Model: This model predicts path loss when the transmitting and receiving antenna are at LOS. The received signal has two components, one is the LOS part and the other component is the signal reflected from the ground. The path loss (dB) is evaluated in [4] and expressed in (17).

$$TR_{free} = \begin{cases} PL_{free}, & d < d_c \\ RL_{free}, & d \geq d_c \end{cases} \tag{17}$$

Where RL_{free} is the Ground Reflected ray model given in [4] and is computed by (18).

$$RL_{free} = 20\log_{10}\left(\frac{d^2}{h_T h_R}\right) \tag{18}$$

and parameter d_c is the crossover distance defined in (19)

$$d_c = \frac{4\pi h_T h_R}{\lambda} \tag{19}$$

Here, in expressions (18) and (19) λ is the wavelength, h_T and h_R denotes the heights of source and destination nodes (m) respectively.

Ray-Tracing Model: Ray Tracing model is a general method of path loss modelling and incorporates the effects like reflections, scattering and diffraction. Here, the path loss values are calculated considering the paths being direct propagation, ground reflection and dielectric property of ice [4].

Log Distance Path Loss Model: This model is utilized when multipath effects and variation in terrain are taken. The expression (dB) is expressed in [4] and given in (20).

$$LD_{free} = L(d_o) + 10\,\mathrm{n}\log_{10}\left(\frac{d}{d_o}\right) + \chi_\sigma \tag{20}$$

here, n is path loss exponent showing rate by which signal received diminishes with distance, $L(d_o)$ is path loss (dB) at reference distance, d_o. χ_σ represents mean Gaussian random variable having standard deviation σ(dB), explaining shadowing effects.

Expression for power received P_r at the destination node is obtained from [18] and is expressed by (21). This expression is also utilized to find the non-cooperative power.

$$P_r = \frac{Energy \times C}{d^{Pathloss}} \tag{21}$$

Where, C is constant derived from [18] and given by Eq. (22).

$$C = G_t G_r \left\lfloor \frac{c}{4\pi f_c} \right\rfloor^2 \tag{22}$$

Here, G_t, G_r denotes antenna gains at transmitter and receiver nodes respectively, f_c denotes carrier frequency, c denotes speed of light.

For simulation and comparitive analysis of results, path loss values have been evaluated using Two ray path loss model, Ray Tracing path loss model and Log Distance path loss model. Received power has been evaluated using expressions (16) and (21) for Non-Cooperative scenario and expressions (13) and (21) for evaluation of cooperative scenario.

3 Simulation and Numerical Results

Method simulation has been performed using MATLAB. Non-Homogenous WSN scenario is assumed. Cooperative communication is implemented using Amplify and Forward relaying protocol. The analysis is done under Log Normal Shadow Fading channel scenario. The modulation technique used is Bipolar Stage Shift Keying (BPSK) and noise considered is Additive White Gaussian Noise (AWGN). The graphs have been plotted taking ideal values of the parameters.

In Fig. 2 distance vs power for direct transmission non-cooperative scenario models of Measured, Log Distance, Two-ray and Ray Tracing are considered. The antenna heights taken are 0.25 m, 1 m and 1.5 m respectively in Figs. 2(a), (b) and (c) respectively. It is evident from the figure that as distance increases, power received at the destination node decreases. The received power further decreases with increase in the antenna height. Higher the antenna height more is the attenuation suffered. It is further observed that for different antenna heights, Empirical log-distance models exhibit the closest values to the measured values. Hence, the Empirical log distance model outperforms other models considered as its evaluated values are closest to the measured values.

In Fig. 3 distance vs power for cooperative scenario models of Measured, Log Distance, Two-ray and Ray Tracing are considered. The antenna heights taken are 0.25 m, 1 m and 1.5 m respectively in Figs. 3(a), (b) and (c) respectively. It is observed that with increase in distance, power received at the destination node decreases. The received power further decreases with increase in the antenna height. However, it is evident from the plots that the cooperative power shows an improvement compared to non-cooperative power. Hence, implementing cooperative communication enhances the performance of WSN. Also, it is observed that for different antenna heights, Empirical log-distance model displays the closest values to the measured values, outperforming other models considered.

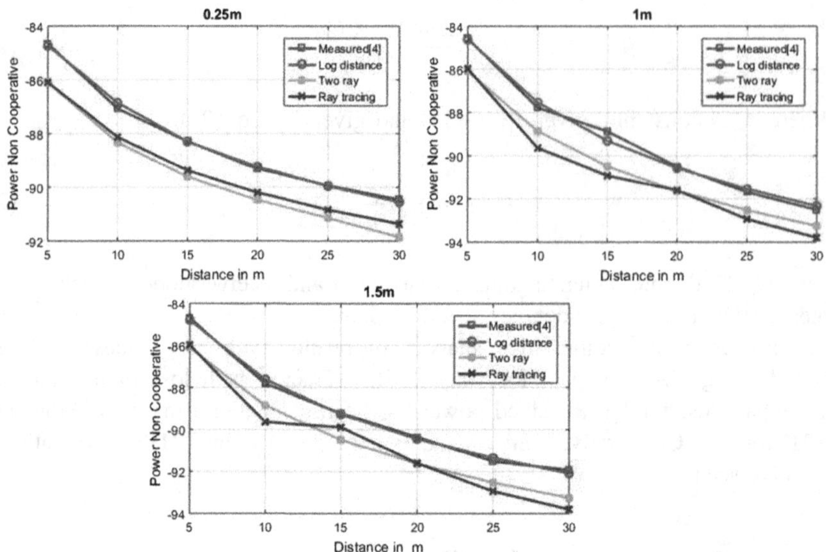

Fig. 2. Distance vs Power (non-cooperative) for different antenna heights

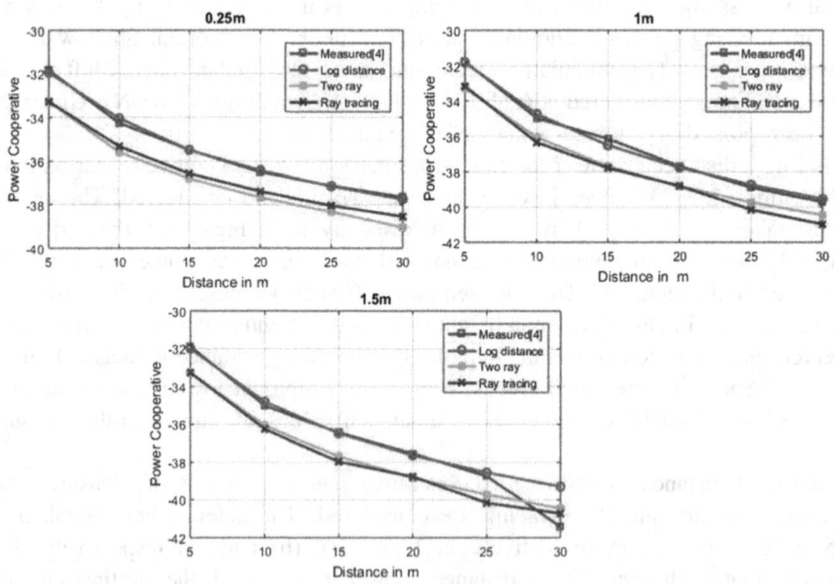

Fig. 3. Distance vs Power (Cooperative) for different antenna heights

In Fig. 4 path loss vs power for direct transmission non-cooperative scenario models of Measured, Log Distance, Two-ray and Ray Tracing are considered. The antenna heights taken are 0.25 m, 1 m and 1.5 m respectively in Figs. 4(a), (b) and (c) respectively. Here, with increase in path loss, power received at the destination node decreases. The received power further decreases with increase in the antenna height. Also, it is observed that for different antenna heights, Empirical log-distance models exhibit the closest values to the measured values.

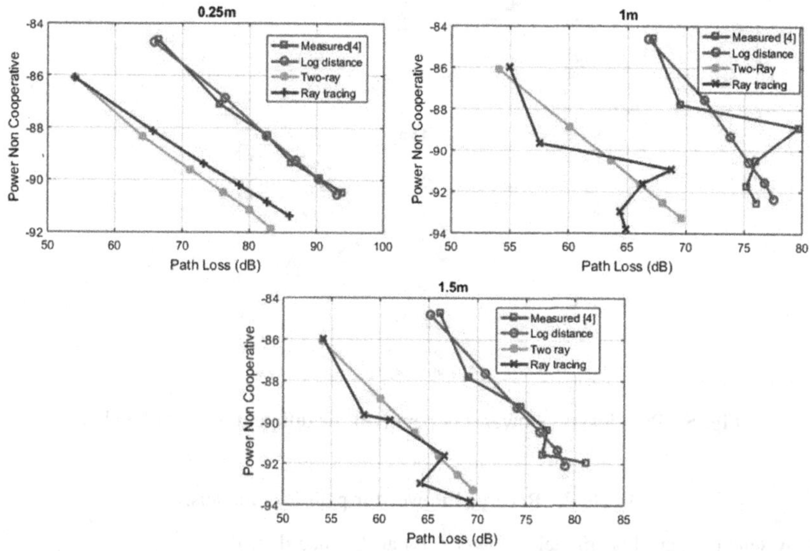

Fig. 4. Path loss vs Power (non-cooperative) for different antenna heights

In Fig. 5 path loss vs power for cooperative scenario models of Measured, Log Distance, Two-ray and Ray Tracing are considered. The antenna heights taken are 0.25 m, 1 m and 1.5 m respectively in Figs. 5(a), (b) and (c) respectively. With increase in path loss, power received at destination node decreases. The received power further decreases with increase in the antenna height. However, it is evident from the plots that the cooperative power shows an improvement compared to non-cooperative power. Hence, implementing cooperative communication enhances the reliability of WSN. Also, it is observed that for different antenna heights, Empirical log-distance models exhibit the nearest values to the measured values, outperforming other models considered.

Table 2 is a comparitive analysis of received power for cooperative and non-cooperative relaying for the models taken. Tabulated results are created considering three distances (d = 10 m, 20 m and 30 m) between the source node and the destination nodes and three antenna heights 0.25 m, 1 m and 1.5 m are taken respectively. The paper is summarized below by stating some critical conclusions.

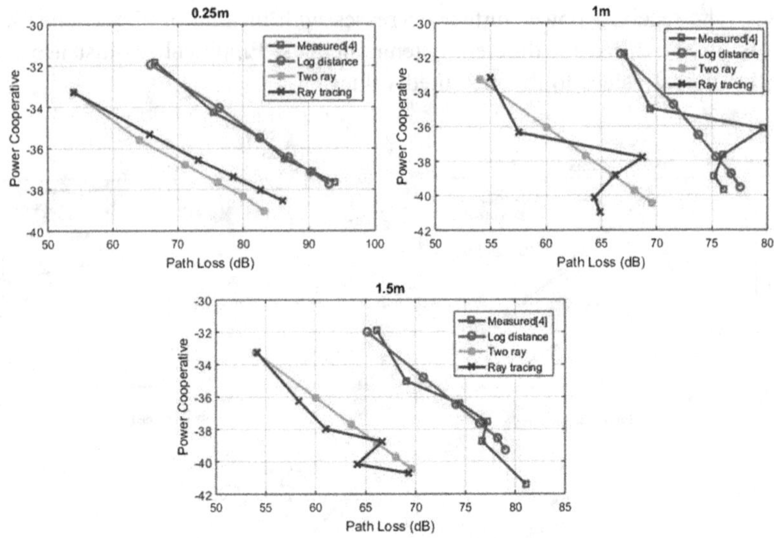

Fig. 5. Path loss vs Power (cooperative) for different antenna heights

Table 2. Received Power for path loss models.

S.no	Antenna height(m)	Path loss models	Power (dB) at distance d (m)					
			Non-cooperative scenario			Cooperative scenario		
			10	20	30	10	20	30
1.	0.25	Measurement	−87.07	−−89.30	−90.46	−34.27	−36.50	−37.65
		Log distance	−86.85	−89.23	90.55	−34.05	−36.42	−37.74
		Ray-Tracing	−88.12	−90.20	−91.36	−35.32	−37.39	−38.55
		Two ray	−88.33	−90.47	−91.85	−35.58	−37.66	−39.05
2.	0.5	Measurement	−87.77	−90.49	−92.50	−34.97	−37.68	−39.70
		Log distance	−87.53	−90.56	−92.33	−34.74	−37.75	−39.53
		Ray-Tracing	−89.68	−91.60	−93.79	−36.34	−38.80	−40.96
		Two ray	−88.85	−91.63	−93.25	−36.05	−38.82	−40.44
3.	1	Measurement	−87.82	−90.35	−91.93	−35.01	−37.54	−41.42
		Log distance	−87.62	−90.44	−92.08	−34.80	−37.63	−39.28
		Ray-Tracing	−89.68	−91.60	−93.79	−36.24	−38.76	−40.71
		Two ray	−88.85	−91.63	−93.25	−36.05	−38.82	−40.44

4 Conclusion

Effectiveness of cooperative communication in a WSN deployed in a snowy environment with adjustable height antennas has been analyzed in this work. The variable height antennas of sensor nodes have been considered for performance evaluation as the signals from short heighted nodes might get buried under the snow, not giving appropriate values of the pathloss. Received power has been taken as the critical parameter for the comparative analysis of cooperative and non-cooperative scenario. Amplify and Forward cooperative relaying scenario has been implemented to derive expressions through mathematical modelling for computing power at the destination node. Path loss models have been used to compute received power. It is seen that the gain offered by the relay relates to bandwidth, rate and amplification factor. Critical comparison of the received power for the cooperative scenario shows an improvement of about 56% compared to the non- cooperative scenario. Computation has been done for variable antenna height scenario. It is also observed that increasing the antenna height reduces the received **power**. Further, it has also been observed that Empirzzzical Log-distance model is a better performing pathloss model. The error percentage of Empirical Log distance model with respect to measured results is 0.02% in the case of non-cooperative scenario and 1.87% in the case of cooperative scenario.

References

1. Aseri, T.C., Kumar, D., Patel, R.B.: EEHC: energy efficient heterogeneous clustered scheme for wireless sensor networks. Comput. Commun. **32**(4), 662–667 (2009)
2. Wang, J., Ghosh, R.K., Das, S.K.: A survey on sensor localization. J. Control Theory Appl. **8**(1), 2–11 (2010)
3. Xiao, X., Zhang, R.: Study of immune-based intrusion detection technology in wireless sensor networks. Arab. J. Sci. Eng. **42**(8), 3159–3174 (2017)
4. Cheffena, M., Mohamed, M.: Empirical path loss models for wireless sensor network deployment in snowy environments. IEEE Antennas Wirel. Propag. Lett. **16**, 2877–2880 (2017)
5. Malhotra, J., Rani, S., Talwar, R.: Energy efficient chain based cooperative routing protocol for WSN. Appl. Soft Comput. **35**, 386–397 (2015)
6. Butcharoen, S., Pirak, C.: An adaptive cooperative protocol for multi-hop wireless networks. In: 2015 17th International Conference on Advanced Communication Technology (ICACT), Seoul, pp. 620–635 (2015)
7. Garcia, M., et al.: Saving energy and improving communications using cooperative group-based wireless sensor networks. Telecommun. Syst. **52**(4), 2489–2502 (2013)
8. Hedley, M., Sathyan, T.: Evaluation of algorithms for cooperative localization in wireless sensor networks. In: 2009 IEEE 20th International Symposium on Personal, Indoor and Mobile Radio Communications, Tokyo, pp. 1898–1902 (2009)
9. Gupta, S.H., Singh, R.K., Sharan, S.N.: Performance analysis of coded cooperation and space time cooperation with multiple relays in Nakagami-*m* fading. In: Gavrilova, M.L., Tan, C.J.K., Saeed, K., Chaki, N., Shaikh, S.H. (eds.) Transactions on Computational Science XXV. LNCS, vol. 9030, pp. 172–185. Springer, Heidelberg (2015). https://doi.org/10.1007/978-3-662-47074-9_10
10. Zarifi, K., Zaidi, S., Affes, S., Ghrayeb, A.: A distributed amplify-and-forward beamforming technique in wireless sensor networks. IEEE Trans. Signal Process. **59**(8), 3657–3674 (2011)
11. Matyjas, J.D., Mo, Z., Su, W.: Amplify and forward relaying protocol design with optimum power and time allocation. In: MILCOM 2016 - 2016 IEEE Military Communications Conference, Baltimore, MD, pp. 412–417 (2016)

12. Ouyang, S., Xiao, H.: Power allocation for a hybrid decode–amplify–forward cooperative communication system with two source–destination Pairs under outage probability constraint. IEEE Syst. J. **9**(3), 797–804 (2015)
13. Kurt, S., Tavli, B.: Path-loss modeling for wireless sensor networks: a review of models and comparative evaluations. IEEE Antennas Propag. Mag. **59**(1), 18–37 (2017)
14. Aldosary, A., Alsayyari, A., Kostanic, I., Otero, C.E.: An empirical path loss model for wireless sensor network deployment in a dense tree environment. In: 2017 IEEE Sensors Applications Symposium (SAS), Glassboro, NJ, pp. 1–6 (2017)
15. Kennedy, S., Stewart, J., Stewart, R.: Internet of things—propagation modelling for precision agriculture applications. In: 2017 Wireless Telecommunications Symposium (WTS), Chicago, IL, pp. 1–8 (2017)
16. Comeau, F., Phillips, W.J., Robertson, W., Sivakumar, S.: A clustered wireless sensor network model based on log-distance path loss. In: 6th Annual Communication Networks and Services Research Conference Halifax, pp. 366–372 (2008)
17. Iqbal, Z., Kim, K., Lee, H.N.: A cooperative wireless sensor network for indoor industrial monitoring. IEEE Trans. Ind. Inform. **13**(2), 482–491 (2017)
18. Bai, F., Bharati, S., Thanayankizil, L.V., Zhuang, W.: Link-layer cooperation based on distributed TDMA MAC for vehicular networks. IEEE Trans. Veh. Technol. **66**(7), 6415–6427 (2017)
19. Devarajan, N., Gupta, S.H.: Implementation and analysis of different path loss models for cooperative communication in a wireless sensor network. In: Tiwari, S., Trivedi, M.C., Mishra, K.K., Misra, A.K., Kumar, K.K. (eds.) Smart Innovations in Communication and Computational Sciences. AISC, vol. 851, pp. 227–236. Springer, Singapore (2019). https://doi.org/10.1007/978-981-13-2414-7_22

A Built-In Circuit for Self-reconfiguring Mesh-Connected Processor Arrays with Spares on Diagonal

Itsuo Takanami[1] and Masaru Fukushi[2(✉)]

[1] Department of Technology, Yamaguchi University, Tokiwadai 2-16-1, Ube 755-0097, Japan
iftakanami@comet.ocn.ne.jp

[2] Graduate School of Sciences and Technology for Innovation, Yamaguchi University, Tokiwadai 2-16-1, Ube 755-0097, Japan
mfukushi@yamaguchi-u.ac.jp

Abstract. This paper presents a built-in self-reconfiguring system for mesh-connected processor arrays where faulty processing elements are compensated for by spare processing elements located on the diagonal. First, an algorithm for reconfiguring the arrays with faulty processing elements is presented. The reliability of the system is analyzed by simulation and compared with that of arrays having spare processing elements on one side. The result shows that under the condition of the same number of spares, the former is fairly higher than the latter. Next, a logical circuit realizing self-reconfiguration by the algorithm is described. The circuit controls interconnections of processing elements and its hardware overhead is shown to be quite small (i.e. less than ten logic gates for each processing element). The proposed system is effective for the case where each processing element is fairly reliable and the small number of spares is sufficient for retaining a high reliability. It is also effective in enhancing the run-time reliability of a processor array in mission critical systems where first self-reconfiguration is required without an external host computer or manual maintenance operations.

Keywords: Fault-tolerance · Mesh array · Spare on diagonal · Self-repair · Built-in circuit

Acronyms:

PA Processor array
PE Processing element
SRS Self-reconfiguration system
DSR Direct spare replacement
STS Single track shifting
RC Reconfigurability condition
RA Reconfiguration algorithm

M. L. Gavrilova and C. J. K. Tan (Eds.): Trans. on Comput. Sci. XXXIV, LNCS 11820, pp. 109–135, 2019.
https://doi.org/10.1007/978-3-662-59958-7_7

1 Introduction

In resent years, there has been a rapidly growing interest in processing many kinds of vast amount of information in real-time and near-real-time. The demand for strengthening computation power will never stop, and it is increasing day by day. For these needs, how to realize high-speed and massively parallel computer has been studied in the literature. A mesh-connected PA is a kind of form of massively parallel computers. Mesh-connected PAs consisting of hundreds of PEs offers a regular and modular structure, a small wiring length between PEs, and a high scalability, thus very suitable for most signal and image processing algorithms.

One of the most important and fundamental issues that must be addressed for such PAs is defect/fault tolerance. If a single PE fails to perform its assigned task correctly, due to some defects/faults, the entire computation will result in failure. On the other hand, as VLSI technology has developed, the realization of parallel computing systems using multi-chip module (MCM) e.g., [1], wafer scale integration (WSI) e.g., [2] or network-on-chip (NoC) e.g., [3] has been considered so as to enhance the computation and communication performance, decrease energy consumption and sizes, and so on. In such a realization, entire or significant parts of PEs and interconnections among them are implemented on a single chip or wafer. Therefore, the yield and/or reliability of the system may become drastically low if no strategy is employed for coping with defects and faults.

To restore the correct computation capabilities of PAs, it must be reconfigured appropriately so that the defective PEs are eliminated from the computation paths, and the working PEs maintain correct logical connectivities between themselves. Various strategies to reconfigure a faulty physical system into a fault-free target logical system are described in the literature, e.g., [4–9]. These strategies are in redundancy approach. In contrast, there are those in degradation approach in which no spare PEs are designated in advance and a reconfiguration is done by avoiding faulty PEs with keeping the array structure, e.g., [10–13]. Some of these techniques employ very powerful reconfiguring systems that can repair a faulty PA with almost certainty, even in the presence of clusters of multiple faults. However, the key limitation of these techniques is that they are executed in software programs to run on an external host computer and they cannot be designed and implemented efficiently within a PA chip as dedicated circuits. If a faulty PA can be reconfigured by a built-in circuit, the system down time of the PA is significantly reduced. Furthermore, the PA will become more reliable when it is used in such an environment where the fault information cannot be monitored externally through the boundary pins of the chip and manual maintenance operations are difficult.

As far as we know, the first attempts to develop SRSs were made by Negrini et al. [14] and Sami and Stefanelli [15]. The design approach of their repair control scheme begins with a heuristic algorithm that involves only some local reconnection operations.

Takanami et al. also developed the automatic SRSs [16–19] for mesh-connected PAs using single track switches proposed by Kung et al. [4]. The arrays they dealt with have two rows and two columns of spares, i.e. four linear arrays of spares around a mesh array to cope with faults of low reliable PEs. Besides the approaches as the above, Lin et al. [20] proposed the fault-tolerant path router with built-in self-test/self-diagnosis and fault-isolation circuits for 2D-mesh based chip multiprocessor systems. Collet et al. [21] also proposed the chip self-organization and fault tolerance at the architectural level to improve dependable continuous operation of multicore arrays in massively defective nanotechnologies, where the architectural self-organization results from the conjunction of self-diagnosis and self-disconnection mechanism, plus self-discovery of routers to maintain the communication in the array.

On the other hand, if PEs are fairly reliable, it is expected that the smaller number of spares will be sufficient for retaining the reliability of an array so high and additional control schemes as well as control circuits will become simple. From this expectation, Takanami et al. [22–24] proposed SRSs for mesh arrays with spares on one row and one column. In these schemes, $2N$ spares are used for arrays with size of $N \times N$. However, the case where the number of spares is less, that is, N, has not been discussed as far as we know.

This paper is a revised and extended version of [25] and deals with the case that the number of spares is N for an $N \times N$ array. As an arrangement of N spares for this case, a simple one is to locate them on a side of an array, which is shown in Fig. 1(a) and called singe-side spare scheme (SS scheme) in which a faulty PE is replaced by the spare on the same row. Another to be proposed here is to locate them on the diagonal of an array, which is called a diagonal spare scheme (DS scheme). Since how faulty PEs are replaced in SS scheme is simple, in Sect. 2, we describe how faulty PEs are replaced in DS scheme. Two methods which are called DSR and STS methods are introduced.

In Sect. 3, reconfiguration probabilities of DS scheme (DSR and STS methods) are shown by simulation, comparing with those of SS scheme. It is seen that the probabilities of DS scheme are fairly higher than those of SS scheme.

In Sect. 4, a logical circuit realizing the algorithm for STS method is presented. Then, the compatibility between the behaviors of the circuit and the algorithm are proven. The proof is a little complicated because the spares are on the diagonal. The circuit is so simple that its hardware overhead is small. The proposed system is effective for the case where each PE is fairly reliable and the small number of spares will be sufficient for retaining the reliability of an array so high under small hardware overhead. It is also effective in enhancing especially the run-time reliability of a PA.

In Sect. 5, we discuss the array reliabilities when arrays have $2N$ spares, the layout of PEs on the diagonal, the failure of the built-in circuit and the improvement of the RA.

Section 6 is the conclusion.

2 DS Scheme

Figure 1(b) shows how PEs with spares are arranged in an array for DS scheme. If a PE p_{ij} located at the i-th row and j-th column is faulty, it is compensated for by either the spare s_i or s_j. For this arrangement, several compensation methods may be considered according to how an array is reconfigured using spares. Here, two methods are introduced. One is to directly replace a faulty PE with a spare, which is called DSR method. Another is to replace a faulty PE with a spare by shifting PEs as in [4,8,9], which is called STS method. In such DSR method, the replacement of faulty PEs by spares on diagonal could be characterized using a bipartite graph in graph theory. Then, As seen in the following, DSR method realizes higher reconfiguration probabilities than STS method. However, DSR method has a disadvantage in that phsical distances among logically adjacent PEs after reconfiguration (compensation) are proportional to the array sizes. On the other hand, STS method has an advantage in that physical distances among logically adjacent PEs after reconfiguration are bounded by a constant. So, here we focus on STS method.

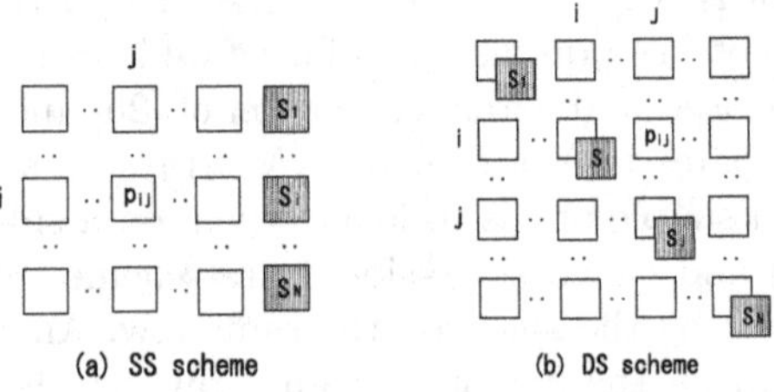

Fig. 1. Arrangement of PEs and spares

Notation:

- If a faulty PE is replaced with a spare by shifting as in STS method, it is said to be repaired, otherwise unrepaired.
- For a set of faulty PEs which is often called a fault pattern, if all the faults in the set can be repaired at the same time, the fault pattern as well as the array with the fault pattern is often said to be repairable, otherwise unrepairable.

Figure 2 shows the network structure for STS. A single track runs between adjacent rows and between adjacent columns, and there is a switch at the intersection point of a track and a link connecting PEs. The reconfiguration is done as follows, using those tracks and switches.

A faulty PE is bypassed and replaced by it's adjacent healthy PE, which in turn is replaced by the next adjacent healthy PE, and so on. This replacement is repeated until a healthy spare PE is used in the end. Here, each PE has four internal switches SWs as shown in Fig. 3 which are switched according to whether it is healthy or faulty, and a faulty PE is bypassed as shown in (b). The reconfiguration process defines a compensation path (shortly written as c-path). Note that the c-path is straight and continuous, and must not pass other

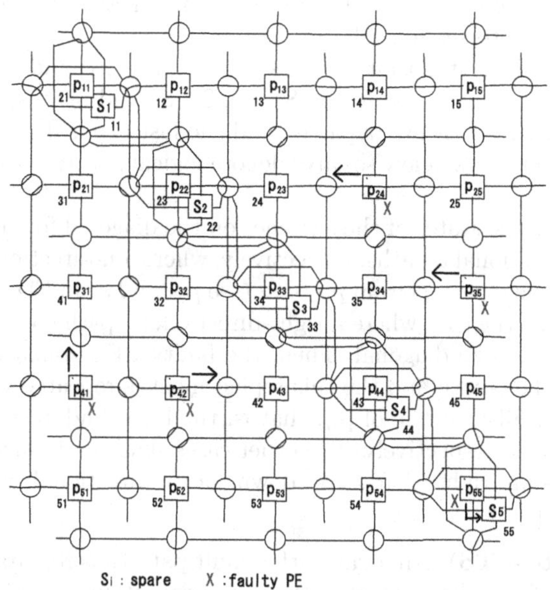

Fig. 2. Network structure for STS method and an illustration of states of swithces to compensate for a fault pattern where ×s denote faulty PEs. Arrows show the directions of compensation for faulty PEs

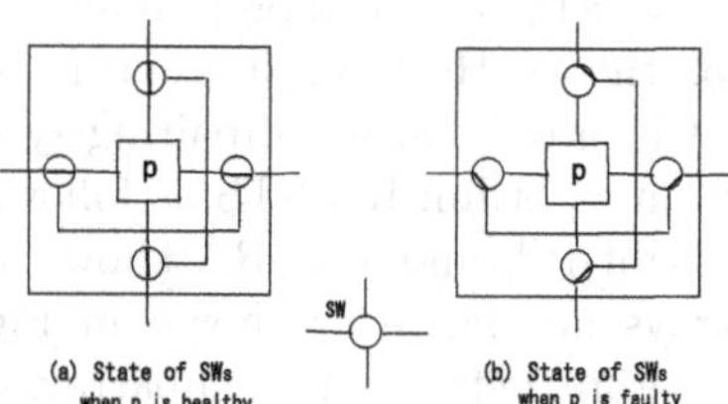

Fig. 3. SWs are switched as (a) and (b) according to whether PE is healthy and faulty, respectively

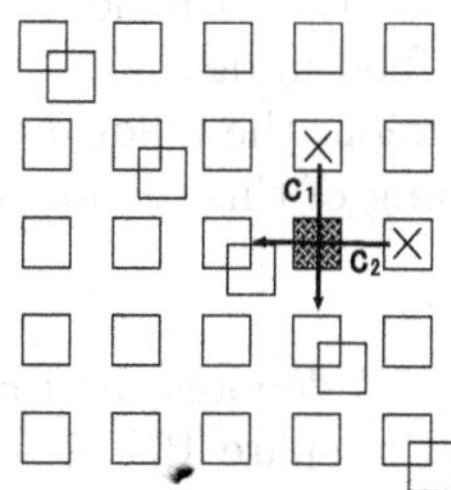

Fig. 4. c-paths c_1 and c_2 in intersection relation.

faulty PEs. Then, if c-paths c_1 and c_2 have the same PE as shown in Fig. 4, they are called to be in intersection relation.

In Fig. 2, it is illustrated that the PEs p_{24}, p_{35}, p_{41}, p_{42} and p_{55} marked with $\times$s are faulty and compensated for by S_2, S_3, S_1, S_4 and S_5, respectively, where the former four faulty PEs are bypassed and the last one is directly replaced by S_5. Further, the numbers below squares denote logical addresses of the PEs after compensation.

Figure 5 shows the states of the switches on the diagonal for right-, left-, up-, down- and on-directional c-paths, respectively, where 'on-directional' means that p_{ii} is faulty and the c-path is from p_{ii} to s_i (see p_{55} and s_5 in Fig. 2). Figure 5(a), for example, shows the case where a right-directional c-path comes from a faulty PE to a spare S_i on the diagonal. Then, the faulty PE is replaced by adjacent healthy PEs on the c-path and the this replacement is repeated until the c-path reaches S_i and S_i fills the role of p_{ii}, that is, the logical addresses of p_{ii} and S_i is $(i, i-1)$ and (i, i), respectively. The other cases shall be similarly understood.

Since c-paths are right-, left-, up-, down- or on-directional, we have the following property [4].

Property 1 (RC-STS). An array with a fault pattern is repairable if and only if there is a set of c-paths S which satisfies the condition

1. S contains a c-path for each nonspare faulty PE, and
2. there is no intersection among the c-paths in S. □

A strategy to get a set of c-paths S for an array with a fault pattern, which satisifies the condition of RC-STS, can also be formalized as matching and independent problems in graph theory. However, it seems to be difficult to realize the strategy in hardware, that is, a built-in self-rapairing system. Hence, we present another method given by an algotihm RA-STS as follows.

RA-STS consists of Right-RA and Left-RA shown below. Given an array, first, two triangle subarrays are made as shown in Fig. 6 in which PEs and spares on the diagonal overlap between two subarrays. They are often called the right- and left-triangle subarrays, and denoted as RT- and LT-subarrays, respectivley. Then, Right-RA is applied to the RT-subarray with spares to compensate for faults in it. Next, Left-RA is applied to the LT-subarray with spares not used in Right-RA to compensate for faults in the LT-subarray. Note that Left-RA is obtained by exchanging the variables i and j in Right-RA with small modifications, and so, not described in detail.

From convenience of explanation, the following notation are used instead of the notation used in the beginning of this section such as p_{ij}.

Notation:

- PE(i, j) $(1 \leq i \leq N, 1 \leq j \leq N)$ denotes the PE at location (i, j) in a PA.
- $S(i)$ $(1 \leq i \leq N)$ denotes the spare PE at location (i, i), that is, on the diagonal of the array.
- p_{ij} (s_i) is the faulty state of PE(i, j) $(S(i))$ where $p_{ij}(s_i) = 1$ and $= 0$ mean that PE(i, j) (spare $S(i)$) is faulty and healthy, respectively.

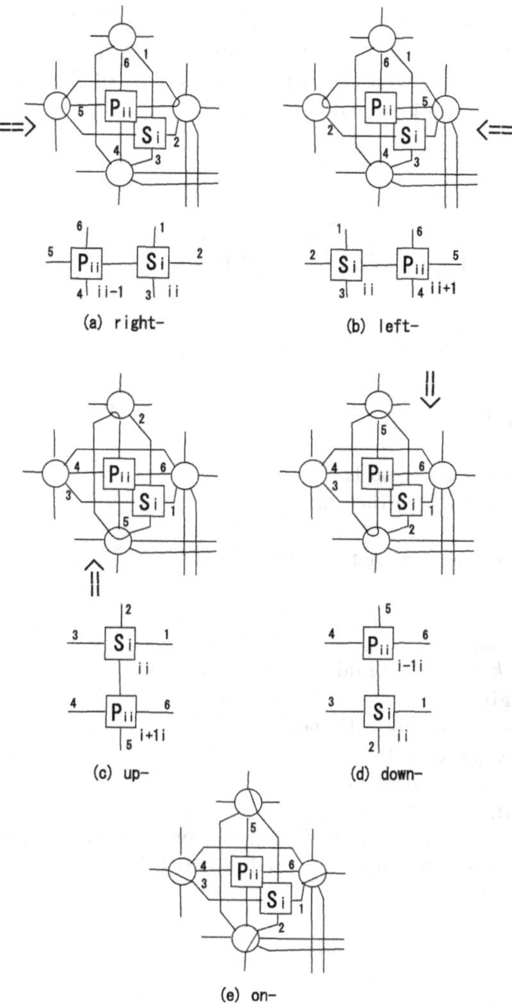

Fig. 5. States of the switches on the diagonal for right-, left-, up- and down-directional c-paths whose directions are shown as arrows, respectively

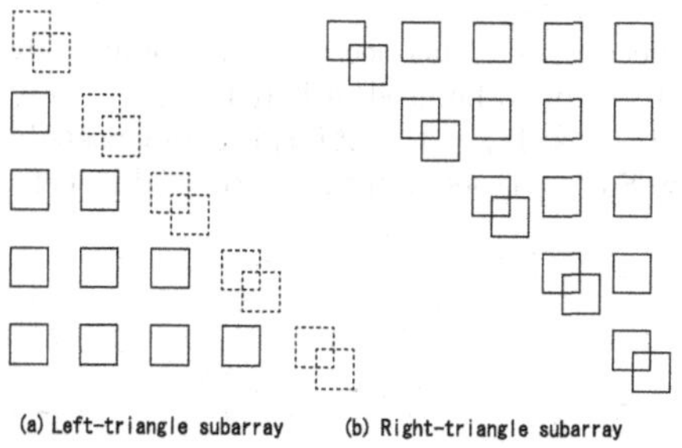

Fig. 6. Left and right triangle subarrays.

- A matrix $F = (f_{ij})(1 \le i \le N, 0 \le j \le N)$ is called a fault pattern where $f_{ij} = p_{ij}$ for $1 \le i \le N$ and $1 \le j \le N$ and $f_{i0} = s_i$.
- $D = (d_{ij})$ be a matrix with the same size as that of P where $P = (p_{ij})$.
- sfg and flg are variables used as flags.

RA-STS

[Right-RA]
begin
(1) set d_{ij} to 0 for i, j $(1 \le i \le N)$, $(1 \le j \le N)$;

(2) set *sfg* to 1;
for $j = N$ to 1 do
begin
(3) set *flg* to s_j;
for $i = j$ to 1 do
begin
(4) if $d_{ij} = 2$ then set *flg* to 1;
(5) if ($p_{ij} = 1$ and *flg*=1) then do
begin
(6) if $(i = j)$ then set *sfg* to 0; stop;
else
begin
(7) set d_{ij} to 2;
for $k = j - 1$ to i do
begin
(8) if ($s_i = 1$ or $p_{ik} = 1$) then
set *sfg* to 0; stop;
(9) else set d_{ik} to 2;
end;
(9-a) set s_i to 1; because S(i) has been used to compensate for PE(i, j),
s_i is set to 1 not to be used in Left-RA
end;
end;
if ($p_{ij} = 1$ and *flg*=0) then do
begin
(10) set *flg* to 1;
for $k = i$ to j do
begin
(11) set d_{kj} to 1;
end;
(11-a) set s_j to 1; because S(j) has been used to compensate for PE(i, j),
s_j is set to 1 not to be used in Left-RA
(11-b) set p_{jj} to 0; because If $i = j$, PE(j, j) is compensated for, otherwise, PE(j, j) is healthy. So, p_{jj} is set to 0 to be accessed in Left-RA
end;
end;
end;
end;

[Left-RA]

This is derived by exchanging the row and column indices in Right-RA without the processes (9-a), (11-a) and (11-b) □

From Right-RA, we can see the following. If $p_{ij} = 1$ and $flg = 1$, a sequence of $d_{ik} =$ 2's ($k \leq j$) is generated in the left direction from PE(i, j) toward the spare at the location (i, i) on the diagonal at the processes (7) and (9) except the case of $i = j$. This sequence which reaches the spare is called 2-sequence from PE(i, j) which is denoted as 2-seq(i, j). In this case, since the spare S(i) has been used to compensate for the faulty PE(i, j), s_i is set to 1 so that the spare S(i) is not used in Left-RA. Similarly, if $p_{ij} = 1$ and $flg = 0$, S(j) is healthy and used to compensate for the faulty PE(i, j). Then, a sequence of $d_{kj} =$ 1's is generated downward from PE(i, j) toward the spare at the location (j, j) on the diagonal at the process (11). This sequence is called 1-sequence from PE(i, j) which is denoted as 1-seq(i, j). In this case, since the spare S(j) has been used to compensate for the faulty PE(i, j), s_j is set to 1 so that the spare S(j) is not used in Left-RA. Further, if $i = j$, PE(j, j) has been compensated for, otherwise PE(j, j) is healthy. So, p_{jj} is set to 0 to be accessed in Left-RA.

The process in Right-RA is outlined as follows.

1. Starting at the rightmost column, that is, the N-th column, we try to find a downward c-path from a selected faulty PE to a spare on the diagonal where PE's are sequentially selected upward from the diagonal. In this case, a sequence of 1's from the selected PE is generated.
2. Otherwise, we try to find a leftward c-path to a spare on the diagonal. In this case, a sequence of 2's from the selected PE is generated.
3. If the above processes are successful for all PEs on the column, that is, $sfg = 1$, the similar processes are executed on the column next to the left, and so on.
4. If the above processes are at last successfully executed on the left-most column, Right-RA ends with $sfg = 1$ and the fault pattern in the RT-subarray is judged to be repairable. Then, a faulty PE(i, j) is repaired by downward (leftward) shifting according to 1-seq(i, j) (2-seq(i, j)).
5. Otherwise, the algorithm ends with $sfg = 0$ and the fault pattern in the RT-subarray is judged to be unrepairable.

The process in Left-RA is executed as in Right-RA but from the bottom row, (i.e. the N-th row) toward the top row, (i.e. the first row).

Then we have the following properties for repairability for an array.

Property 2 (Right-RC). A fault pattern in an RT-subarray is repairable in the meaning of Property 1 if and only if Right-RA ends with $sfg = 1$. If it ends with $sfg = 1$, the set of 1- and 2-sequences is a set of c-paths which satisfies RC-STS described in Property 1.

Proof: See Appendix A. □

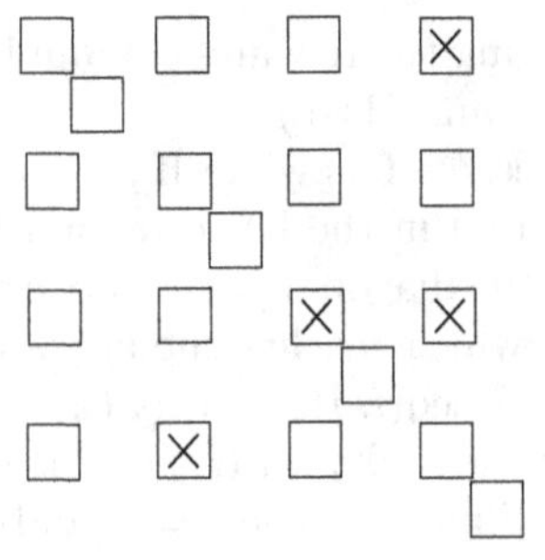

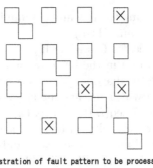

(a) An illustration of fault pattern to be processed in RA-STS

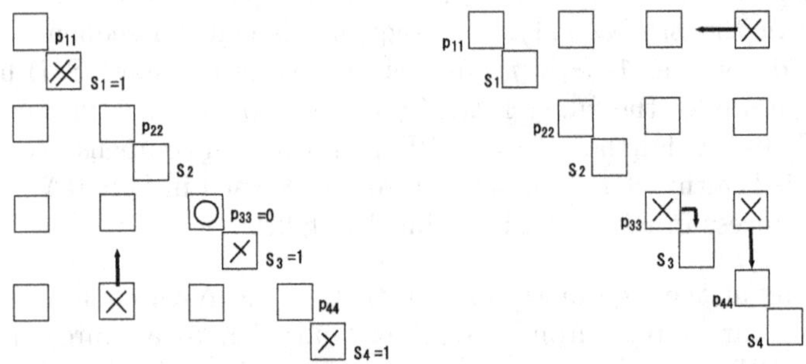

(b) The subpattern of (a) to be processed in Left-RA where [figure] [figure] and [figure] are those which are modified corresponding to (9-a), (11-a) and (11-b), respectively

(c) The subpattern to be processed in Right-RA

Fig. 7. An iilustration of a fault pattern where arrow lines indicate the directions of c-paths. (c) is the right subpattern to be processed in Right-RA. (b) is the left subpattern to be processed in Left-RA, which is modified at (9-a), (11-a) and (11-b) in Right-RA.

The following is obtained from Property 2.

Property 3

1. An array with a fault pattern is repairable in the meaning of Property 1 if RA-STS applied to the fault pattern ends with $sfg = 1$. If it ends with $sfg = 1$, the set of 1- and 2-sequences is a set of c-paths which satisfies the condtion in Property 1.
2. If Right-RA applied to a fault pattern ends with $sfg = 0$, there is no set of c-paths for the array with the fault pattern which satisfies the condition in Property 1. □

3 Array Reliabilities

Figures 8 and 9 show the array reliabilities (ARs) with logical sizes $N \times N$ for $N = 4$ and 8, respectively, under the condition that each PE may be healthy with equal probability p, where $AY(k) = \frac{N_{rep}(k)}{N_{pat}(k)}$, $AR(p) = \sum_{k=0}^{N_s} {}_{Na}C_k \cdot AY(k) \cdot p^{N_a - k}(1-p)^k$, $N_{rep}(k)$, $N_{pat}(k)$, N_s, and N_a are the number of fault patterns which have k faulty PEs and are judged to be repairable, the number of examined fault patterns which have k faulty PEs, the number of spare PEs, and the number of all PEs, respectively.

Further, $*x*$-DSR and -STS are the array reliabilites for DSR and STS methods, respectively, which are derived by Monte Carlo simulation. $*x*$-SS is that for SS scheme which is calculated by the equation $[p^{N+1} + (N+1)p^N(1-p)]^N$.

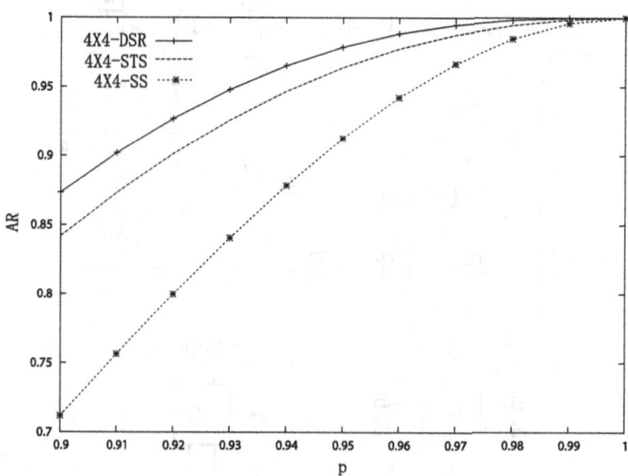

Fig. 8. Relation between AR and p for 4×4 arrays

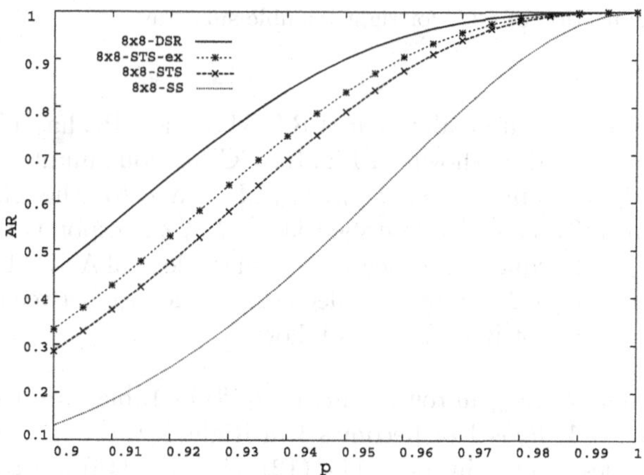

Fig. 9. Relation between AR and p for 8×8 arrays

From the results, it can be seen that the ARs for DSR and STS methods are fairly higher than that for SS scheme. ARs of PAs with larger sizes will be discussed in Sect. 5. 8 × 8-STS-ex in Fig. 9 will also be explained in Sect. 5.

4 SRS

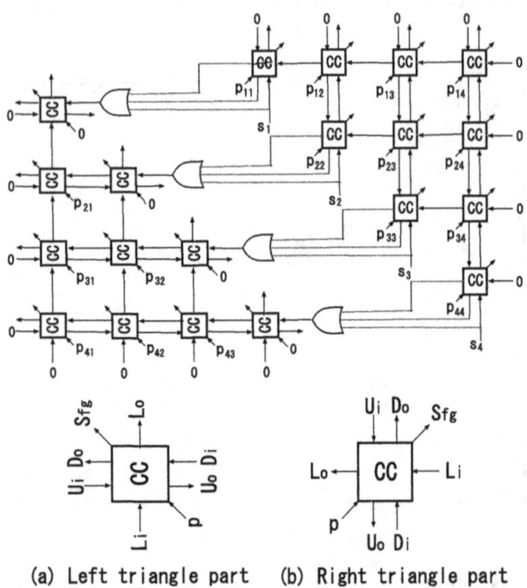

Fig. 10. A hardware realization of RA-STS. (a) Connection of *CC*s for left triangle subarray. (b) Connection of *CC*s for right trianble subarray.

Figure 10 shows a hardware realization of RA-STS. Each PE has a logical circuit CC, which is connected as shown in Fig. 10. CC has four input signals p, D_{in}, L_{in} and U_{in} and four ouput signals D_{out}, L_{out}, U_{out} and S_{fg} the relations among which are given in Table 1. As described later, D_{in} corresponds to *flg*, L_{out} to 2-sequence, U_{out} to 1-sequence, respectively, and the logical AND of S_{fg}'s from all CC's corresponds to *sfg*. The more detailed correspondence between the behavior of CC and the process of RA-STS is as follows.

1. That the values of D_{out} in rows from (1) to (8) in Table 1 are 1's corresponds to that if $p_{ij} = 1$, *flg* is 1 or becomes 1 in Right-RA.
2. That the values of D_{out} in rows (11), (12), (15) and (16) are 1's corresponds to the process (4) in Right-RA.

Table 1. Truth table of CC ($*$ means *don't care*) where S_{fg}^{nd} and S_{fg}^{d} are S_{fg}'s for nondiagonal and diagonal, respectively

	p	D_{in}	L_{in}	U_{in}	D_{out}	L_{out}	U_{out}	S_{fg}^{nd}	S_{fg}^{d}
(1)	1	0	0	0	1	0	1	1	1
(2)	1	0	0	1	1	$*$	$*$	$*$	1
(3)	1	0	1	0	1	$*$	$*$	0	0
(4)	1	0	1	1	1	$*$	$*$	$*$	0
(5)	1	1	0	0	1	1	0	1	0
(6)	1	1	0	1	1	$*$	$*$	$*$	0
(7)	1	1	1	0	1	$*$	$*$	0	0
(8)	1	1	1	1	1	$*$	$*$	$*$	0
(9)	0	0	0	0	0	0	0	1	1
(10)	0	0	0	1	0	0	1	1	1
(11)	0	0	1	0	1	1	0	1	1
(12)	0	0	1	1	1	$*$	$*$	$*$	1
(13)	0	1	0	0	1	0	0	1	1
(14)	0	1	0	1	1	$*$	$*$	$*$	1
(15)	0	1	1	0	1	1	0	1	0
(16)	0	1	1	1	1	$*$	$*$	$*$	0

3. That the values of D_{out} in rows (13) and (14) are 1's and that the values of D_{out} in rows (9) and (10) are 0's are due to that if $p = 0$, none of from the processes (4) to (11) in Right-RA are applied and *flg* does not change.
4. The first row in Table 1 corresponds to the process (11) in Right-RA and causes the origin of 1-sequence.
5. Rows (3) and (7) correspond to the process (8) in Right-RA and hence, $S_{fg} = 0$.
6. Row (5) corresponds to the process (7) in Right-RA and causes the origin of 2-sequence.
7. Row (9) corresponds to the case where a healthy PE is not on a c-path.
8. Row (10) corresponds to the case where 1-sequence passes a healthy PE.
9. Row (11) corresponds to the case where 2-sequence passes a healthy PE.
10. Row (12) corresponds to the case where 1- and 2-sequences intersect each other. However, it can be shown that this case never occurs in a steady state.
11. Row (13) corresponds to the case where PE is healthy, and none of from the processes (4) to (9) in Right-RA is applied.
12. Row (15) corresponds to the case where 2-sequence passes a healthy PE.

From Table 1, we have the following logical equations using *don't care* terms.

$$D_{out} = p + D_{in} + L_{in} \tag{1}$$

$$L_{out} = L_{in} + p \cdot D_{in} \tag{2}$$

$$U_{out} = U_{in} + p \cdot \overline{D}_{in} \tag{3}$$

$$\overline{S_{fg}^{nd}} = p \cdot L_{in} \tag{4}$$

$$\overline{S_{fg}^{d}} = p \cdot (L_{in} + D_{in}) + L_{in} \cdot D_{in} \tag{5}$$

From the equations above, we can see that the logical circuit of a CC is so simple that the numbers of gates and pins are less than ten and no flip-flop is used.

In the following, we will describe that the behavior of CCs' network (shortly written as CC-net) shown in Fig. 10 is compatible with that of RA-STS, that is, Properties 5, 6 and 7. Property 5 is used for checking whether an array with faulty PE's is repairable or not. Property 6 shows the correspondence between 2-sequences and sequences of L_{out}'s. Property 7 shows the correspondence between 1-sequences and sequences of U_{out}'s.

For convenience of explanation, the terminal names and the signals from/in the terminals are often identically used unless confused, and the signals from/in CC of PE(x, y) are denoted with index '(x, y)'.

Note that $L_{in}(x, N) = 0$ for any $x (1 \leq x \leq N)$ and $U_{in}(1, y) = 0$ for any $y (1 \leq y \leq N)$. Further, let $L_{in}(y+1, y) = D_{in}(y+1, y) = 0$ and $p_{y+1y} = s_y$. Then, $D_{out}(y+1, y) = D_{in}(y, y) = s_y$.

Since the CC-net shown in Fig. 10(b) behaves in asynchronous mode, we will first prove that the signals in the network become stable for any fixed inputs.

Lemma 1. Let t_d denote signal propagation time in CC. For any column of CC's whose size is $k \times 1$, the signals in the column become stable within $(2k-1) \cdot t_d$ after external inputs to the column are fixed.

Proof: Note that the external inputs to the y-th column are $D_{in}(y, y) (= s_y)$, $L_{in}(x, y) (1 \leq x \leq y)$, $p_{xy} (1 \leq x \leq y)$ and $U_{in}(1, y)$, and the outputs from the y-th column go to the left, but not to the right. Now, let the external inputs to the column be fixed. We will prove by induction on k. The statement clearly holds for the case of $k = 1$. Let $k \geq 2$. Since from Eq. (1) $D_{out}(k, k)$ depends only on the external inputs $D_{in}(k, k) (= s_k)$, p_{kk} and $L_{in}(k, k)$ inputs, $D_{in}(k-1, k) (= D_{out}(k, k))$ becomes stable in t_d. Then, since the inputs to CC's on the k-th column which consists of rows except the k-th row are fixed, from the hypothesis of induction the signals in all CC's on k-th column become stable within $(2(k-1)-1) \cdot t_d$. After that, since $U_{in}(k, y) = U_{out}(k-1, y)$, $U_{out}(k, y)$ is fixed in t_d. In total, the signals become stable within $(2k-1) \cdot t_d$. □

Since external signals on a column come from its right neighbors, we have the following property.

Property 4. For the CC-net for the RT-subarray of an array of logical size $N \times N$, the signals become stable within $N^2 \cdot t_d$ for any fault pattern. □

In the following, from Property 4 we consider that the signals have become stable. Now, we show the following properties which are closely related to hardware realization of RA-STS.

Property 5. The RT-subarray of an array with a fault pattern is repairable if and only if all the output signals S_{fg}'s from CC's are 1's.

The proof: see Appendix B-1. □

Now, we discuss how to switch the connections among PE's if the logical AND of all the output signals S_{fg}'s from CC's is 1. Hence, in the following, we assume that a repairable fault pattern is given, that is, the logical AND of all the output signals S_{fg}'s from CC's is 1. In order to switch the connections among PE's correctly, from Property 2 it is necessary and sufficient to know the correspondences between 1- or 2-sequences in RA-STS and the signals in the CC-net. Such correspondences are given by the following properties.

Property 6. $d_{xy} = 2$ if and only if $L_{out}(x, y) = 1$.

Proof: See Appendix B-2. □

Property 7. $d_{xy} = 1$ if and only if $U_{out}(x, y) = 1$.

Proof: See Appendix B-3. □

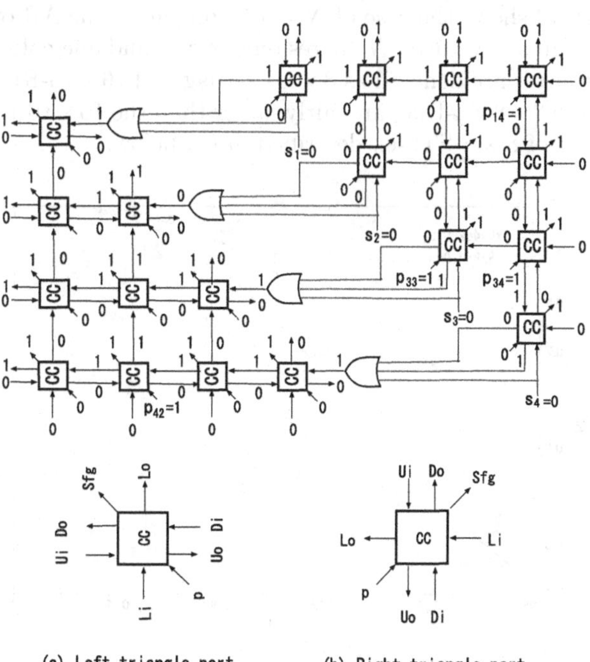

Fig. 11. An illustration of signals in CC-net for a 4 × 4 array with faults shown in Fig. 7.

From the above, there are one-to-one correspondences between 1-sequences and sequences of U_{out}'s, and between 2-sequences and sequences of L_{out}'s. Then, from Property 2, we could give a truth table defining the states of switches, using U_{out}'s and L_{out}'s. However, it will be omitted since it is not so difficult task.

Figure 11 illustrates signals in CC-net for a 4×4 array with the faults shown in Fig. 7. The signal 1 from L_{out} of $CC(1,1)$ in the right triangle part to the OR-gate connected to D_i of $CC(1,1)$ in the left triangle part corresponds to (9-a) in Right-RA. Similarly, 1's from U_{out} of $CC(3,3)$ in the right triangle part to the OR-gate connected to D_i of $CC(3,3)$ in the left triangle part and from that of U_{out} of $CC(4,4)$ to L_{in} of $CC(4,4)$ correspond to (11-a) in Right-RA. $p_{33} = 0$ of $CC(3,3)$ in the left triangle part instead of $p_{33} = 1$ in the right triangle part corresponds to (11-b).

5 Discussion

(1) We have investigated the case that the number of spares is N for an $N \times N$ array from the reason mentioned in Introduction. Of course, it is necessary to increase the number of spares to cope with the case that each PE is not so reliable. A method to increase the number of spares using DS scheme is to divide a given array into subarrays each of which DS scheme is applied to. For example, an $N \times N$ array is divided into four subarrays each with the size of $\frac{N}{2} \times \frac{N}{2}$ and DS scheme is applied to each subarray. Then, the $N \times N$ array has $2N$ spares and the AR of the array is calculated by the 4-th power of that of the subarray. Figure 12 shows the case of $N = 16$, comparing the AR of the array in DS with that of an array with $2N$ spares on one row and one column where the former and the latter are denoted as $8 \times 8 \times 4$-Diag and 16×16-RL, respectively. It is seen that the $8 \times 8 \times 4$-Diag are fairly larger than the 16×16-RL. $8 \times 8 \times 4$ex is the 4-th power of 8×8-STS-ex described in (4) below.

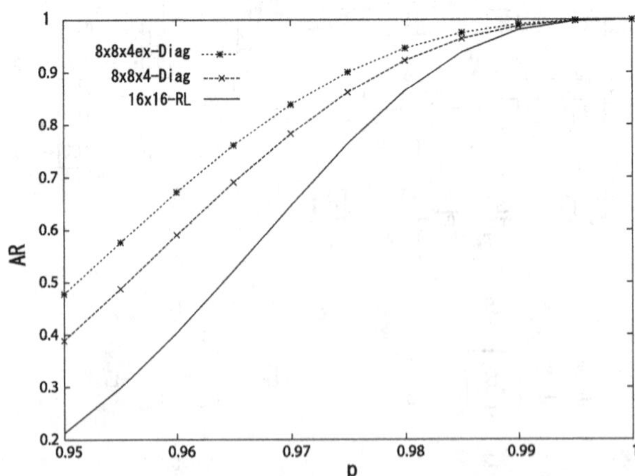

Fig. 12. Relation between ARs and p's for 16×16 arrays where $8 \times 8 \times 4$-Diag is the 4th power of AR of 8×8 array in DS and 16×16-RL is that of 16×16 array with spares on one row and one column

(2) Concerning the layout of PE(i, i) and S(i) on diagonal, as can be seen in Fig. 2, it is not difficult to put them in two dimension and install them together in a module though the size of the module may become larger than that of a PE on nondiagonal. Further, only the failure of the connections in the module must be considered since those of PE(i, i) and S(i) are already considered. However, since the hardware complexity of a PE is usually much larger than that of the connection, the failure of the connection is considered to be negligible (see [24]).

(3) The failure of the built-in circuit can also be considered to be negligible since the hardware complexity of the circuit is usually much less than that of a PE (see [24]).

(4) Property 5 has given a sufficient condition that an array with a fault pattern is repairable. It is hoped that a necessary and sufficient condtion will be given to achieve the highest reliability in the DS scheme using STS method. However, even if it is given, it seems that the hardware realization, that is, a built-in circuit for it, will become complicated. So, here we present an extended algorithm to achieve higer reliability as Right-RA(h). It is the modification of Right-RA where i and j in Right-RA are exchanged and small changes are made according to the exchange. If more concretely stated,

- Exchanged i and j.
- "for $j = N$ to 1" in (2) of Right-RA is changed to "for $i = 1$ to N".
- "for $i = j$ to 1" in (3) is to "for $j = i$ to N".
- "for $k = j - 1$ to i" in (7) is to "for $k = i + 1$ to j".
- "if ($s_i = 1$ or $p_{ik} = 1$) then" in (8) is to "if ($s_j = 1$ or $p_{kj} = 1$) then".
- "else set d_{ik} to 2" in (9) is to "else set d_{kj} to 2".
- "set d_{kj} to 1; " in (11) is to "set d_{ik} to 1;"

From the similarity between Right-RA and Right-RA(h), we have the following property.

Property 8. *Right-RA(h) ends with $sfg = 1$ if and only if Right-RA ends with $sfg = 1$.* □

Then, the extended algorithm RA-STS-ex is as follows.

[RA-STS-ex]

1. Execute Right-RA in RA-STS. If it ends with $sfg = 1$, goto (3).
2. (else it ends with $sfg = 0$) A given fault pattern is not repairable and goto (6)
3. Execute Left-RA. If it ends with $sfg = 1$, the fault pattern is judged to be repairable and goto (6)
4. Execute Right-RA(h) followed by Left-RA. If it ends with $sfg = 1$, the fault pattern is judged to be repairable and goto (6)

5. (else it ends with $sfg = 0$) The fault pattern is judged to be unrepairable.
6. The algorithm ends.

The AR simulated is shown in Fig. 9 as the curve labeled 8 × 8-STS-ex. It is seen that the AR obtained by RA-STS-ex increases, comparing with that by RA-STS.

Now, we describe a hardware realization of RA-STS-ex. CC for Right-RA shown in Fig. 13(a) are the same as those in the right triangle part in Fig. 10(b). CC for Right-RA(h) shown in Fig. 13(b) is that for CC in (a) turned clockwise by 90°, and L_i and L_o are exchanged. Then, Fig. 13(a) and (b) are realized as shown in Figs. 14 and 15, by only switching sw's around CC, respectively.

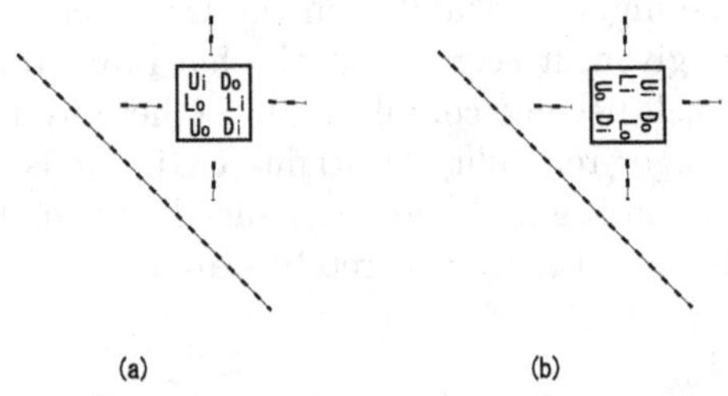

Fig. 13. (a) CC's in the hardware realization of Right RA. (b) CC's in that of Right-RA(h)

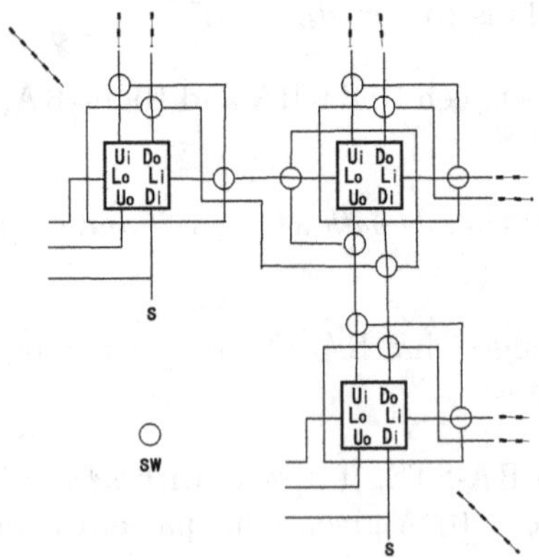

Fig. 14. State of sw's in CC's on executing Right-RA in RA-STS -ex

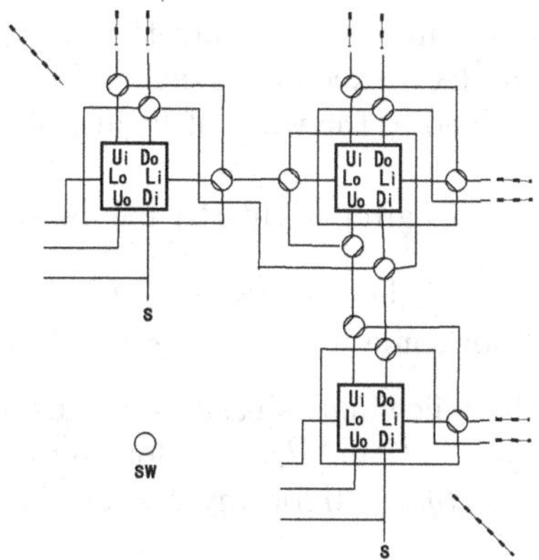

Fig. 15. State of sw's in CC's on executing Right-RA(h) in RA-STS -ex

6 Conclusion

Toward the realization of SRSs, we have presented a RA for $N \times N$ mesh-connected PAs with N spares on diagonal using STS and a built-in digital circuit to realize the algorithm. The circuit is so simple that each CC consists of several gates (i.e. less than ten logic gates for each PE) and no flip-flop is used. The proposed system is effective for the case where each processing element is fairly reliable and the small number of spares is sufficient for retaining a high reliability. It is also effective in enhancing the run-time reliability of a processor array in mission critical systems where first self-reconfiguration is required without an external host computer or manual maintenance operations.

Acknowledgment. The research is in part supported by Yamaguchi University Fund.

Appendix A: The proof of Property 2

Notation:

- **RC-STS**: reconfigurability condition described in Property 1.
- **RA-STS**: the total algorithm consisting of Right-RA and Left-RA.
- **Right-RA**: the sub-algorithm of RA-STS for right triangle subarray.
- **Left-RA**: the sub-algorithm of RA-STS for left triangle subarray.

Property 2 is proved by a sequence of the following Proposition and Lemmas.

Proposition 1. In processing a column of an array, if *flg* changes from 0 to 1 at a row, it does not change from 1 to 0 at the succeeding rows. $sfg = 1$ at the begining of a process, and the process ends with $sfg = 0$ if *sfg* changes from 1 to 0, otherwise with $sfg = 1$.

Proof: This is easily shown from Right-RA. □

Notation: (1) The process that 'for-statement' at $j = j_0(1 \leq j_0 \leq N)$ and $i = i_0(1 \leq i_0 \leq j)$ in Right-RA is executed is denoted as "for-process at (i_0, j_0)".

(2) $flg_{i_0 j_0}(1 \leq i_0 \leq j_0)$ denotes the value of *flg* just after for-process at (i_0, j_0) is executed.

(3) $d_{i_0 j_0}$s denote the values after Right-RA has been executed.

Lemma 2. For any $j(N \geq j \geq 1)$ and any $i(j \geq i \geq 1)$, $flg_{ij} = 1$ if and only if $s_j = 1$, $p_{\hat{i}j} = 1$ or $d_{\hat{i}j} = 2$ for some $\hat{i}(j \geq \hat{i} \geq i \geq 1)$.

Proof: From Right-RA the followingt is easily seen. If $flg_{ij} = 0$, $flg_{i'j} = 0$ for any $i' \geq i$. Then $s_j = 0$, $p_{i'j} = 0$ and $d_{i'j} = 0$ for any $i' \geq i$. If $s_j = 0$, $p_{i'j} = 0$ and $d_{i'j} = 0$ for any $i' \geq i$, $flg_{i'j} = 0$ for any $i' \geq i$. □

Lemma 3. 1. 1-seq(i, j) never passes other faulty PEs than PE(i, j).
2. 2-seq(i, j) never passes other faulty PEs than PE(i, j).

Proof: (1) 1-seq(i, j) is generated downward at (11) in for-process at (i, j) from a faulty PE(i, j). Hence, $p_{ij} = 1$ and $flg_{(i+1)j} = 0$. Therefore, from Lemma 2 $p_{\hat{i}j} = 0$ for all $\hat{i}(j \geq \hat{i} \geq i + 1)$ and $s_j = 0$, and this means that the 1-seq(i, j) never passes other faulty PEs (including the spare $S(j)$) than PE(i, j).

(2) Since a 2-sequence is generated at (7) and (9) in for-process at (i, j) from a faulty PE(i, j), and it reaches a spare, it is clear that it never passes other faulty PEs than PE(i, j). □

Lemma 4. 2-sequences neither intersect nor overlap each other.

Proof: It is clear that they never intersect since they go only in the left direction. We suppose that there exist 2-seq(i, j) and 2-seq$(i', j')((i, j) \neq (i', j'))$ which overlap. Then $p_{ij} = p_{i'j'} = 1$ and $i = i'$. Without loss of generality, it can be assumed that $j' > j$. Then *sfg* becomes 0 at (8) in for-process at (i, j') and a sequence of 2's from PE(i, j') can not reach the spare S(i). This contradicts that there exist 2-seq(i, j) and 2-seq(i', j') which overlap. □

Lemma 5. 1-sequences neither intersect nor overlap each other. Furthermore, 1-sequences and 2-sequences never intersect each other.

Proof: It is clear that no 1-sequences intersect each other since they go only downward. We suppose that there exist 1-seq(i, j) and 1-seq(i', j') $((i, j) \neq (i', j'))$ which overlap. Then $p_{ij} = p_{i'j'} = 1$ and $j = j'$ since a 1-sequence is generated at (11). Without loss of generality, $i' > i$ can be assumed. Then 1-seq(i, j) passes the faulty PE(i', j'), which contradicts Lemma 3. Next, we suppose that there exist 1-seq(i, j) and 2-seq(i', j') which intersect. Then $i' \geq i$ and $j' \geq j$, and from the flow of execution in Right-RA 1-seq(i, j) is generated after 2-seq(i', j') is generated. If $i = i'$, 2-seq(i', j') passes the faulty PE(i, j), which contradicts Lemma 3. If $i' > i$, the process (4) in for-process at (i', j) is executed since $d_{i'j} = 2$, and $flg_{i'j} = 1$. Hence, from Proposition 1, $flg_{ij} = 1$. Therefore, the process (10) is not executed in for-process at (i, j), and 1-seq(i, j) is not generated. This contradicts that there exist 1-seq(i, j) and 1-seq(i', j') $((i, j) \neq (i', j'))$ which overlap. □

Lemma 6. Let $p_{ij} = 1$, that is, PE(i,j) be faulty. If $flg_{(i+1)j} = 1$, the downward c-path (if it exists) from PE(i,j) is not included in any set of c-paths which satisfies RC-STS described in Property 1.

Proof: Let $p_{ij} = flg_{(i+1)j} = 1$. Then, from Lemma 2 $s_j = 1$, $p_{\hat{i}j} = 1$, or $d_{\hat{i}j} = 2$ for some $\hat{i}(j \geq \hat{i} \geq i+1)$. We will prove by induction on the number of 2-sequences t. If $t = 0$, no d_{uv} is 2 and hence, $s_j = 1$ or $p_{\hat{i}j} = 1$ for some $\hat{i}$ $(j \geq \hat{i} \geq i+1)$, and it is clear that there is not a downward c-path from PE(i,j).

Let $t \geq 1$. For the case where $s_j = 1$ or $p_{\hat{i}j} = 1$ for some $\hat{i}(j \geq \hat{i} \geq i+1)$, the proof is similar as above. Otherwise, $s_j = 0$ and $p_{i'j} = 0$ for any $i'(j \geq i' \geq i+1)$ and $d_{\hat{i}j} = 2$ for some $\hat{i}$ $(j \geq \hat{i} \geq i+1)$. Then there exists 2-seq$(\hat{i}, j')$ for some j' $(j' \geq j)$. Just before the 2-seq$(\hat{i}, j')$ is generated at (5) to (9) in for-process$(\hat{i}, j')$, the number of 2-sequences is less than t, and $d_{\hat{i}j'} = 0$, $p_{\hat{i}j'} = 1$ and $flg_{\hat{i}j'} = 1$. Therefore, since (4) is not executed but (5) is executed, $flg_{\hat{i}+1j'} = 1$, and from the hypothesis of induction, the downward c-path (if it exists) from PE$(\hat{i}, j')$ is not included in any set of c-paths which satisfies RC-STS. This implies that a c-path from faulty PE$(\hat{i}, j')$ must be taken in the left direction. However, the downward c-path from PE(i,j) intersects the left-directional c-path from PE$(\hat{i}, j')$. This implies that the downward c-path (if it exists) from PE(i,j) is not included in any set of c-paths which satisfies RC-STS. □

The proof of Property 2. (**'only if' part**): We assume that the algorithm ends with $sfg = 0$ and suppose that there exists a set S of c-paths which satisfies RC-STS. *sfg* becomes 0 at (6) or (8) in for-process at some (i,j). Then $p_{ij} = 1$ (that is, PE(i,j) is faulty) and $flg = 1$ at (5). If $flg_{(i+1)j} = 0$, $d_{ij} = 2$ at (4) and hence, a 2-seq(i,j') for some j' $(j < j')$ passes the faulty PE(i,j). This contradicts (2) in Lemma 3. Therefore $flg_{(i+1)j} = 1$. Then, from Lemma 6, S must contains the left-directional c-path from PE(i,j). However, there exist faulty PEs in the left direction from PE(i,j) since *sfg* beomes 0, and hence the left-directional c-path for PE(i,j) is also not included in S. This contradicts the property of S. Therefore, there exists no set of c-paths which satisfies RC-STS.

(**'if' part**): We assume that the algorithm ends with $sfg = 1$. If $flg = 1$ at (5) in for-process at (i,j) for $p_{ij} = 1$ (that is, PE(i,j) is faulty), the processes (7) and (9) generate 2-seq(i,j) which reaches the spare on the diagonal in the left-directiion from (i,j). If $flg = 0$ for $p_{ij} = 1$, the processe (11) generates 1-seq(i,j) which reaches the spare on the diagonal in the downward from (i,j). From Lemma 3 these sequences are c-paths from PE(i,j). Furthermore, these sequences from faulty PEs neither intersect nor overlap each other. Therefore, the set of 1- and 2-sequences obtained satisfies RC-STS. □

Appendix B-1: The proof of Property 5

Property 5 is proved by a sequence of the following Lemmas.

The following Lemmas 7, 8 and 9 are easily proved, referring to Eqs. (1), (2) and (3).

Lemma 7. (1) If $D_{out}(x, y) = 1$, for all $x'(x' \leq x) D_{out}(x', y) = 1$.
(2) If $L_{out}(x, y) = 1$, for all $y'(y' \leq y)\ L_{out}(x, y') = 1$.
(3) If $U_{out}(x, y) = 1$, for all $x'(x' \geq x)\ U_{out}(x', y) = 1$. □

Proof; The statements clearly hold from Eqs. (1), (2) and (3). □

Lemma 8. If $L_{out}(x, y) = 1$, $D_{out}(x, y) = 1$.

Proof: The statement is easily proved from Eqs. (1) and (2). □

Lemma 9. If $D_{out}(x, y) = 1$, $U_{in}(x, y) = 0$.

Proof: If $D_{out}(x, y) = 1$, for any $x' < x$ $D_{in}(x', y) = D_{out}(x' + 1, y) = 1$ from Lemma 7, and hence $U_{out}(x', y) = U_{in}(x', y)$ from Eq. (3), and since $U_{in}(1, y) = 0$, $U_{out}(x', y) = 0$ for any $x' < x$. Therefore, $U_{in}(x, y) = U_{out}(x - 1, y) = 0$. □

Lemma 10. For $y(N \geq y \geq 1)$ and $x(y+1 \geq x \geq 1)$, $D_{out}(x, y) = 1$ if and only if for some $x'(y+1 \geq x' \geq x)\ p_{x'y} = 1$ or for some $x''(y \geq x'' \geq x) L_{in}(x'', y) = 1$.

Proof: ('only if' part) We will prove by induction on x, decreasing x. The statement holds for $x = y + 1$ since $p_{y+1y} = D_{out}(y + 1, y)$. Let $x \leq y$ and suppose that the statement is true for any $x'(y + 1 \geq x' > x)$. Let $D_{out}(x, y) = 1$. If $D_{in}(x, y)\ (= D_{out}(x+1, y)) = 1$, from the hypothesis of induction the statement holds. Hence, let $D_{in}(x, y) = 0$. Then, from Eq. (1) $p_{xy} = 1$ or $L_{in}(x, y) = 1$ and hence, the statement holds.

('if' part) We will prove by induction on x, decreasing x. Since $p_{y+1y} = D_{out}(y+1, y)$, the statement holds for $x = y+1$. Suppose that for any $\hat{x}(y+1 \geq \hat{x} > x)$ the statement is true. If for some $x'(y + 1 \geq x' > x)\ p_{x'y} = 1$ or for some $x''(y \geq x'' > x) L_{in}(x'', y) = 1$, from the hypothesis of induction, $D_{out}(x + 1, y) = 1$. Hence, from (1) of Lemma 7 $D_{out}(x, y) = 1$. If $p_{xy} = 1$ or $L_{in}(x, y) = 1$, from Eq. (1) $D_{out}(x, y) = 1$. □

Lemma 11. For $y(N \geq y \geq 1)$ and $x(y \geq x \geq 1)$, $L_{out}(x, y) = 1$ if and only if for some $y'(N \geq y' \geq y) p_{xy'} = 1$ and $D_{in}(x, y') = 1$.

Proof: ('only if' part) Let $L_{out}(x, y) = 1$. Since $L_{in}(y+1, y) = D_{in}(y+1, y) = 0$, we assume $x \leq y$. We will prove by induction on y, decreasing y. Let $y = N$. Since $L_{in}(x, N) = 0$, from Eq. (2) $1 = L_{out}(x, N) = p_{xN} \cdot D_{in}(x, N)$. Hence, the statement holds for $y = N$. Suppose that the statement is true for any $y'(N \geq y' > y \geq 1)$. From Eq. (2) (i) $L_{in}(x, y) = 1$ or (ii) $p_{xy} = 1$ and $D_{in}(x, y) = 1$. If $L_{in}(x, y) = 1$, $L_{in}(x, y) = L_{out}(x, y + 1) = 1$. Hence, from the hypothesis of induction, for some $y'(y' \geq y + 1 > y) p_{xy'} = 1$ and $D_{in}(x, y') = 1$, and the statement holds. If $p_{xy} = 1$ and $D_{in}(x, y) = 1$, the statement clearly holds.

('if' part) From Eq. (2), for some $y' \geq y$ $L_{out}(x, y') = 1$. Hence, from (2) of Lemma 7 $L_{out}(x, y) = 1$. □

Lemma 12. If $L_{out}(x, y) = 1 (y \geq x)$, for some x' and $x''(y \geq x' > x'' \geq x)$ and for some $y'(y' \geq y) p_{x'y'} = p_{x''y'} = 1$, that is, PE($x'y'$) and PE($x''y'$) are faulty.

Proof: We will prove by induction on x and y. Let $y = N$. Since $L_{in}(x, N) = 0$, from Eq. (2) $1 = L_{out}(x, N) = p_{xN} \cdot D_{in}(x, N)$. Hence, $p_{xN} = 1$ and $D_{out}(x + 1,$

$N) = D_{in}(x, N) = 1$. Hence, from Lemma 10, for some $x'(N+1 \geq x' \geq x+1 > x)p_{x'N} = 1$ or for some x'' $(N \geq x'' \geq x+1 > x)$ $L_{in}(x'', N) = 1$. However, since $L_{in}(x'', N) = 0$, $p_{x'N} = 1(x' > x)$, and since $p_{xN} = 1$, the statement holds for any $x(y \geq x \geq 1)$ and $y = N$. Suppose that the statement holds for any $\hat{y}(N \geq \hat{y} > y)$ and any $\hat{x}(\hat{y} \geq \hat{x})$ and let $L_{out}(y, y) = 1$. From Lemma 11 for some $y'(N \geq y' \geq y)$ $p_{yy'} = 1$ and $D_{in}(y, y') = 1$. Hence, $D_{out}(y+1, y') = D_{in}(y, y') = 1$ and from Lemma 10, for some $x'(y'+1 \geq x' \geq y+1)p_{x'y'} = 1$ or some $x''(y' \geq x'' \geq y+1)$ $L_{in}(x'', y') = 1$. If $p_{x'y'} = 1$, the statement holds since $p_{yy'} = 1$. If $L_{in}(x'', y') = 1$, from Eq. (2) $L_{out}(x'', y') = 1(y' > y)$ and hence by the hypothesis of induction, for some x_1 and $x_2(y' \geq x_1 > x_2 \geq x'')$ and some $y''(N \geq y'' \geq y')p_{x_1y''} = p_{x_2y''} = 1$. Hence the statement holds for $x = y$ and y.

Suppose that the statement holds for any y and any $\hat{x}$ $(y \geq \hat{x} > x)$, and let $L_{out}(x, y) = 1$. From Lemma 11, for some $y'(y' \geq y)p_{xy'} = 1$ and $D_{out}(x+1, y') = D_{in}(x, y') = 1$. Then from Lemma 10, for some $x'(y'+1 \geq x' \geq x+1)p_{x'y'} = 1$ or for some $x''(y' \geq x'' \geq x+1)L_{in}(x'', y') = 1$. If $p_{x'y'} = 1$, the statement holds since $x' > x$ and $p_{xy'} = p_{x'y'} = 1$. If $L_{in}(x'', y') = 1$, from Eq. (2) $L_{out}(x'', y') = 1(x'' \geq x+1 > x)$ and by the hypothesis of induction, for some x_1 and $x_2(x_1 > x_2 \geq x'' > x)$ and for some $y''(y'' \geq y' \geq y)p_{x_1y''} = p_{x_2y''} = 1$. Hence, the statement is true. □

Lemma 13. (1) If $d_{xy} = 2$ $(y \geq x)$, for some $y'(y' \geq y)p_{xy'} = 1$ and $flg_{x+1y'} = 1$.
(2) If $d_{xx} = 2$, for some $y'(y' > x)p_{xy'} = 1$ and $flg_{x+1y'} = 1$.
(3) Under the condition that Right-RA ends with $sfg = 1$, $d_{xy} = 2$ if and only if for some $y'(y' \geq y)p_{xy'} = 1$ and $flg_{xy'} = 1$.

Proof: At the beginning of for-process at (x, y),

(i) if $d_{xy} = 2$, for some $y' > yd_{xy'+1} = 0$, $p_{xy'} = 1$ and $(s_{y'} = 1$ or $flg_{x+1y'} = 1)$. Then from Lemma 2 if $s_{y'} = 1$, $flg_{y'y'} = 1$ and hence $flg_{x+1y'} = 1$.
(ii) if $d_{xy} = 0$, $p_{xy} = 1$ and $(s_y = 1$ or $flg_{x+1y} = 1)$. If $y = x$, the for-process ends with $sfg = 0$. Therefore, $y > x$. Then from Lemma 2, if $s_y = 1$, $flg_{yy} = 1$ and hence $flg_{x+1y} = 1$.

From above, (1) and (2) are proved. (3) is easily proved from (1) and (2). □

Lemma 14. If $d_{xy} = 2$, $L_{out}(x, y) = 1$.

Proof: By induction on x and y. If $x = y = N$. Since $d_{NN} = 0$, the statement is proved. Suppose that for all $x'(N \geq x' > x)$ the statement is true, and let $d_{xN} = 2$. From Lemma 13, $p_{xN} = 1$ and $flg_{x+1N} = 1$. From Lemma 2, $s_N(= p_{N+1N}) = 1$, for some $x'(x' \geq x+1 > x)p_{x'N} = 1$ or $d_{x'N} = 2$. If $p_{x'N} = 1$, from Lemma 10 $D_{out}(x', N) = 1$. If $d_{x'N} = 2$, from the hypothesis of induction $L_{out}(x', N) = 1$, and from Lemma 8 $D_{out}(x', N) = 1$. From Lemma 7 $D_{in}(x, N) = D_{out}(x+1, N) = D_{out}(x', N) = 1$. Since $p_{xN} = 1$, from Lemma 11 $L_{out}(x, N) = 1$.

Suppose that for any $\hat{x}(N \geq \hat{x} > x)$ and any $\hat{y}(N \geq \hat{y} > y)$ if $d_{\hat{x}\hat{y}} = 2$, $L_{out}(\hat{x}, \hat{y}) = 1$, and let $d_{xy} = 2(x = y)$. From Lemma 13, for some $y'(y' > y = x)p_{xy'} = 1$ and $flg_{x+1y'} = 1$. Then from Lemma 2 $s_{y'} = 1$, $p_{x''y'} = 1$ or

$d_{x''y'} = 2$ for some $x''(y' \geq x'' \geq x)$. if $s_{y'} = 1$, $D_{in}(y', y') = s_{y'} = 1$, and then $D_{out}(y', y') = 1$ and $D_{out}(x'', y') = 1$. If $p_{x''y'} = 1$, from Eq. (1) $D_{out}(x'', y') = 1$. If $d_{x''y'} = 2$, from the hypothesis of induction $L_{out}(x'', y') = 1$, and then from Lemma 8 $D_{out}(x'', y') = 1$. For either case, $D_{out}(x+1, y') = D_{in}(x, y') = 1$. Then since $p_{xy'} = 1$, from Eq. (2) $L_{out}(x, y') = 1$, and hence $L_{out}(x, y) = 1$ $(x = y)$.

Suppose that for any $\hat{x}(N \geq \hat{x} > x)$ if $d_{\hat{x}y} = 2$ $L_{out}(\hat{x}, y) = 1$ and let $d_{xy} = 2$ $(\hat{x} > x)$. Then from Lemma 13 for some $y'(y' \geq y)p_{xy'} = 1$ and $flg_{x+1y'} = 1$. Similarly to the above, $L_{out}(x, y') = 1$, and $L_{out}(x, y) = 1$. □

Lemma 15 ('only if' part of Property 5). If $S_{fg}(x, y) = 0$ for some $(x, y)(N \geq y \geq 1, y \geq x \geq 1)$, Right-RA ends with $sfg = 0$.

Proof: Let $y_0 = \max_y\{y|S_{fg}(x, y) = 0$ for some $x(y \geq x)\}$ and let $x_0 = \min_x\{x|S_{fg}(x, y_0) = 0\}$. Let $x_0 \leq y_0$. Then, from Eq. (4) $p_{x_0y_0} = 1$ and $1 = L_{in}(x_0, y_0) = L_{out}(x_0, y_0 + 1)$. Then from Lemma 11, for some $y'(N \geq y' \geq y_0 + 1)p_{x_0y'} = 1$ and $D_{in}(x_0, y') = D_{out}(x_0 + 1, y') = 1$. Hence, from Lemma 10, for some $x'(y' + 1 \geq x' \geq x_0 + 1)p_{x'y'} = 1$ or for some $x''(y' \geq x'' \geq x_0 + 1)L_{in}(x'', y') = 1$. If $p_{x'y'} = 1$, $p_{x_0y_0} = p_{x_0y'} = p_{x'y'} = 1$ for $y' > y_0$ and $x' > x_0$, and hence there is no c-path from faulty PE(x_0, y') from Lemma 3, that is, Right-RA ends by returning $sfg = 0$ from Property 2. If $L_{in}(x'', y') = 1$, $L_{out}(x'', y') = 1$ from Eq. (2) and hence from Lemma 12 for some x_1, $x_2(x_1 > x_2 \geq x'' > x_0)$ and some $y''(y'' \geq y' > y_0)p_{x_1y''} = p_{x_2y''} = 1$. Therefore, c-paths from PE(x_2, y'') and PE(x_0, y') are only admissible in the left and downward directions, respectively. However, They cross each other (see the figure below). Hence, there are no c-path for them and hence Right-RA ends by returning $sfg = 0$ from Property 2. □

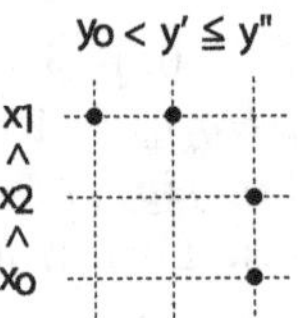

PEs at positions marked with black dots are faulty.

Lemma 16 ('if' part of Property 5). If Right-RA ends with $sfg = 0$, $S_{fg}(x, y) = 0$ for some (x, y).

Proof: If Right-RA ends with $sfg = 0$ in for-process at (x, y), $sfg = 0$ at the process (6) or (8). Hence, If $sfg = 0$ at (6), $x = y$, $p_{yy} = 1$ and $(s_y = 1$ or $d_{yy} = 2)$. If $s_y = 1$, $D_{in}(y, y) = s_y = 1$ and hence from Eq. (4) $\overline{S}_{fg}(y, y) = 1$.

The case that $d_{yy} = d_{xx} = 2$ is impossible since if so, Right-RA would end with $sfg = 0$ in for-process at (x, y') for some $y'(y' > y)$.

Now, if $sfg = 0$ at (8) in for-process at $(x, y)(y > x)$, $p_{xy} = 1$, $d_{xy} = 2$ and $(s_x = 1$ or $p_{xy'} = 1$ for some $y'(y' < y))$. Then from Lemma 14 $L_{out}(x, y) = 1$ and hence from Lemma 7 $L_{in}(x, y') = L_{in}(x, x) = 1$. If $p_{xy'} = 1$, from Eq. (4) $\overline{S}_{fg}(x, y') = 1$. If $s_x = 1$, since $D_{in}(x, x) = s_x = 1$, from Eq. 5 $\overline{S}_{fg}(x, x) = 1$. □

From Lemmas 15, 16 and Properties 2, 5 has been proved.

Appendix B-2: The proof of Property 6

Property 6: ($d_{xy} = 2$ if and only if $L_{out}(x,y) = 1$.

(**'only if' part**) is proved by Lemma 14.

(**'if' part**) Let $L_{out}(x,y) = 1$. From Lemma 11 $p_{xy'} = 1$ and $D_{in}(x,y') = 1$ for some $y'(N \geq y' \geq y)$. By induction on y. Let $y = N$. Then $p_{xN} = 1$ and $D_{in}(x,N) = 1$. Hence, $D_{out}(x+1,N) = 1$. From Lemma 10 and $L_{in}(\hat{x},N) = 0$ for all $\hat{x}$, $p_{x'N} = 1$ for some $x'(x' \geq x+1)$. From Lemma 2 $flg_{x+1N} = 1$ and hence from (3) of Lemma 13 $d_{xN} = 2$, that is, the statement holds for $y = N$. Suppose that the statement is true for any $y''(y < y'' \leq N)$. Since $D_{out}(x+1,y') = D_{in}(x,y') = 1$, from Lemma 10 (i) for some $x'(x' \geq x+1)p_{x'y'} = 1$ or (ii) for some x'' $(y' \geq x'' \geq x+1)L_{in}(x'',y') = 1$. For the case (i), from Lemma 2 $flg_{x+1y'}g = 1$. Since $p_{xy'} = 1$ and $y' \geq y$, From (3) of Lemma 13 $d_{xy} = 2$. For the case (ii), $L_{out}(x'',y'+1) = 1$. From the hypothesis of induction, $d_{x'',y'+1} = 2$. Hence, $d_{x''y'} = 2$. From Lemma 2 $flg_{x+1y'} = 1$. Since $p_{xy'} = 1$, from Lemma (1) of Lemma 13 $d_{xy} = 2$. □

Appendix B-3: The proof of Property 7

Lemma 17. $flg_{xy} = 1$ if and only if $D_{out}(x,y) = 1$.

(**'only if' part**) Let $flg_{xy} = 1$, From Lemma 2 if $s_y = 1$, $s_y = D_{in}(y,y) = D_{out}(y,y) = 1$. Hence, from (1) of Lemma 7, $p_{x'y} = 1$ or $d_{x'y} = 2$ for some $x'(x' \geq x)$. If $p_{x'y} = 1$, from Lemma 10 $D_{out}(x,y) = 1$. If $d_{x'y} = 2$, from Property 6 $L_{out}(x',y) = 1$. From Lemma 8 $D_{out}(x',y) = 1$ and hence, from Lemma 7 $D_{out}(x,y) = 1$.

(**'if' part**) If $D_{out}(x,y) = 1$, from Lemma 10 for some $x'(y+1 \geq x' \geq x)p_{x'y} = 1$ or for some $x''(y \geq x'' \geq x)L_{in}(x'',y) = 1$. If $p_{x'y} = 1$, from Lemma 2 $flg_{xy} = 1$. If $L_{in}(x'',y) = 1$, from Eq. (2) $L_{out}(x'',y) = 1$. Hence, from Property 6 $d_{x''y} = 2$, and from Lemma 2 $flg_{xy} = 1$. □

Lemma 18. $U_{out}(x,y) = 1$ if and only if for some $x'(x' \leq x)$ $p_{x'y} = 1$ and $D_{in}(x',y) = 0$.

Proof: ('only if' part). We will prove by induction on x. Let $x = 1$. Since $U_{in}(1,y) = 0$, from Eq. (3) $U_{out}(1,y) = p_{1y} \cdot \overline{D}_{in}(1,y) = 1$. Hence, $p_{1y} = 1$ and $D_{in}(1,y) = 0$. Suppose that for any $x'(1 \leq x' < x)$ the statement holds. Let $U_{out}(x,y) = 1$. From Eq. (3), (i) $U_{in}(x,y) = 1$ or (ii) $p_{xy} = 1$ and $D_{in}(x,y) = 0$. If $U_{in}(x,y) = 1$, $U_{out}(x-1,y) = U_{in}(x,y) = 1$. Hence, from the hypothesis of induction for some $x'(x' \leq x-1 < x)$ $p_{x'y} = 1$ and $D_{in}(x',y) = 0$. If $p_{xy} = 1$ and $D_{in}(x,y) = 0$, the statement clearly holds.

('if' part). From Eq. (3) $U_{out}(x',y) = 1$ and from (3) of Lemma 7 $U_{out}(x,y) = 1$. □

Property 7: $d_{xy} = 1$ if and only if $U_{out}(x,y) = 1$.

(**'only if' part**) If $d_{xy} = 1$, from Right-RA for some $x'(x' \leq x)p_{x'y} = 1$ and $flg_{x'+1y} = 0$. From Lemma 17, $D_{out}(x'+1,y) = 0$. Since $D_{in}(x',y) = D_{out}(x'+1,y) = 0$ and $p_{x'y} = 1$, from Lemma 18 $U_{out}(x,y) = 1$.

(**'if' part**) If $U_{out}(x, y) = 1$, from Lemma 18 for some $x'(x' \leq x)$ $p_{x'y} = 1$ and $D_{in}(x', y) = 0$. Since $D_{out}(x' + 1, y) = D_{in}(x', y) = 0$, from Lemma 17 $flg_{x'+1y} = 0$. Then, from Lemma 2 $s_y = 0$, for all $x'' \geq x' + 1 p_{x''y} = 0$ and $d_{x''y} = 0$. Then, from Lemma (3) of Lemma 13, $flg_{x'+1} = 0$ and the process (11) in for-statement at (x', y) in Right-RA is executed, and hence $d_{xy} = 1$. □

References

1. Schaper, L.W.: Design of multichip modules. Proc. IEEE **80**(12), 1955–1964 (1992)
2. Okamoto, K.: Importance of wafer bonding for the future hype-miniaturized CMOS devices. ECS Trans. **16**(8), 15–29 (2008)
3. Okamoto, K.: Route packets, not wires: on-chip interconnection networks. In: Proceedings of the 38th Design Automation Conference, pp. 684–689 (2001)
4. Kung, S.Y., Jean, S.N., Chang, C.W.: Fault-tolerant array processors using single-track switches. IEEE Trans. Comput. **38**(4), 501–514 (1989)
5. Mangir, T.E., Avizienis, A.: Fault-tolerant design for VLSI: effect of interconnection requirements on yield improvement of VLSI designs. IEEE Trans. Comput. **c-31**(7), 609–615 (1982)
6. Negrini, R., Sami, M.G., Stefanelli, R.: Fault-Tolerance Through Reconfiguration of VLSI and WSI Arrays. MIT Press Sries in Computer Systems. MIT Press, Cambridge (1989)
7. Koren, I., Singh, A.D.: Fault tolerance in VLSI circuits. IEEE Comput. **23**(7), 73–83 (1990)
8. Roychowdhury, V.P., Bruck, J., Kailath, T.: Efficient algorithms for reconstruction in VLSI/WSI array. IEEE Trans. Comput. **39**(4), 480–489 (1989)
9. Varvarigou, T.A., Roychowdhury, V.P., Kailath, T.: Reconfiguring processor arrays using multiple-track models: the 3 - tracks - 1 - spare - approach. IEEE Trans. Comput. **42**(11), 1281–1293 (1993)
10. Fukushi, M., Fukushima, Y., Horiguchi, S.: A genetic approach for the reconfiguration of degradable processor arrays. In: IEEE 20th International Symposium on Defect and Fault Tolerance in VLSI Systems, pp. 63–71, October 2005
11. Fukushima, Y., Fukushi, M., Horiguchi, S.: An improved reconfiguration method for degradable processor arrays using genetic algorithm. In: IEEE 21st International Symposium on Defect and Fault Tolerance in VLSI Systems, pp. 353–361, October 2006
12. Wu, J., Zhu, L., He, P., Jiang, G.: Reconfigurations for processor arrays with faulty switches and links. In: 15th IEEE/ACM International Symposium on Cluster, Cloud and Grid Computing, pp. 141–148 (2015)
13. Qian, J., Zhou, Z., Gu, T., Zhao, L., Chang, L.: Optimal reconfiguration of high-performance VLSI subarrays with network flow. IEEE Trans. Parallel Distrib. Syst. **27**(12), 3575–3587 (2016)
14. Negrini, R., Sami, M., Stefanelli, R.: Fault tolerance techniues for array structures used in supercomputing. IEEE Comput. **19**(2), 78–87 (1986)
15. Sami, M., Stefanelli, R.: Reconfigurable architectures for VLSI processing arrays. Proc. IEEE **74**, 712–722 (1986)
16. Takanami, I., Kurata, K., Watanabe, T.: A neural algorithm for reconstructing mesh-connected processor arrays using single-track switches. In: International Conference on WSI, pp. 101–110, January 1995

17. Horita, T., Takanami, I.: An efficient method for reconfiguring the $1\frac{1}{2}$ track-switch mesh array. IEICE Trans. Inf. Syst. **E82-D**(12), 1545–1553 (1999)
18. Horita, T., Takanami, I.: Fault tolerant processor arrays based on the $1\frac{1}{2}$-track switches with flexible spare distributions. IEEE Trans. Comput. **49**(6), 542–552 (2000)
19. Horita, T., Takanami, I.: An FPGA implementation of a self-reconfigurable system for the $1\frac{1}{2}$ track-switch 2-D mesh array with PE faults. IEICE Trans. Inf. Syst. **E83-D**(8), 1701–1705 (2000)
20. Lin, S.Y., Shen, W.C., Hsu, C.C., Wu, A.Y.: Fault-tolerant router with built-in self-test/self-diagnosis and fault-isolation circuits for 2D-mesh based chip multiprocessor systems. Int. J. Electr. Eng. **16**(3), 213–222 (2009)
21. Collet, J.H., Zajac, P., Psarakis, M., Gizopoulos. D.: Chip self-organization and fault-tolerance in massively defective multicore arrays. IEEE Trans. Dependable Secur. Comput. **8**(2), 207–217 (2011)
22. Takanami, I.: Self-reconfiguring of $1\frac{1}{2}$-track-switch mesh arrays with spares on one row and one column by simple built-in circuit. IEICE Trans. Inf. Syst. **E87-D**(10), 2318–2328 (2004)
23. Takanami, I., Horita, T.: A built-in circuit for self-repairing mesh-connected processor arrays by direct spare replacement. In: 2012 IEEE 18th Pacific Rim International Symposium on Dependable Computing, pp. 96–104, November 2012
24. Takanami, I., Horita, T., Akiba, M., Terauchi, M., Kanno, T.: A built-in self-repair circuit for restructuring mesh-connected processor arrays by direct spare replacement. In: Gavrilova, M.L., Tan, C.J.K. (eds.) Transactions on Computational Science XXVII. LNCS, vol. 9570, pp. 97–119. Springer, Heidelberg (2016). https://doi.org/10.1007/978-3-662-50412-3_7
25. Takanami, I., Fukushi, M.: A built-in circuit for self-repairing mesh-connected processor arrays with spares on diagonal. In: Proceedings of 22nd Pacific Rim International Symposium on on Dependable Computing, pp. 110–117, January 2017

Author Index

Printed in the United States
By Bookmasters